Statistical Process Control in Industry

MATHEMATICAL MODELLING:
Theory and Applications

VOLUME 5

This series is aimed at publishing work dealing with the definition, development and application of fundamental theory and methodology, computational and algorithmic implementations and comprehensive empirical studies in mathematical modelling. Work on new mathematics inspired by the construction of mathematical models, combining theory and experiment and furthering the understanding of the systems being modelled are particularly welcomed.

Manuscripts to be considered for publication lie within the following, non-exhaustive list of areas: mathematical modelling in engineering, industrial mathematics, control theory, operations research, decision theory, economic modelling, mathematical programming, mathematical system theory, geophysical sciences, climate modelling, environmental processes, mathematical modelling in psychology, political science, sociology and behavioural sciences, mathematical biology, mathematical ecology, image processing, computer vision, artificial intelligence, fuzzy systems, and approximate reasoning, genetic algorithms, neural networks, expert systems, pattern recognition, clustering, chaos and fractals.

Original monographs, comprehensive surveys as well as edited collections will be considered for publication.

The titles published in this series are listed at the end of this volume.

Statistical Process Control in Industry

Implementation and Assurance of SPC

by

R. J. M. M. Does
Institute for Business and Industrial Statistics of the University of Amsterdam, Amsterdam, The Netherlands

K. C. B. Roes
Centre for Biometry Wageningen, Wageningen, The Netherlands

and

A. Trip
Philips Semiconductors, Stadskanaal, The Netherlands

KLUWER ACADEMIC PUBLISHERS
DORDRECHT / BOSTON / LONDON

A C.I.P. Catalogue record for this book is available from the Library of Congress.

ISBN 978-90-481-5170-7

Published by Kluwer Academic Publishers,
P.O. Box 17, 3300 AA Dordrecht, The Netherlands.

Sold and distributed in North, Central and South America
by Kluwer Academic Publishers,
101 Philip Drive, Norwell, MA 02061, U.S.A.

In all other countries, sold and distributed
by Kluwer Academic Publishers,
P.O. Box 322, 3300 AH Dordrecht, The Netherlands.

Printed on acid-free paper

This is a revised translation of the original Dutch work
Statistische Procesbeheersing in Bedrijf,
Kluwer Bedrijfswetenschappen, Deventer © 1996
Translated by the authors.

To my wife Dorine and our sons Kay, Martijn, Jeroen and Sjoerd R.J.M.M.D.

To my wife Sabine and our sons Okko, Jelger and Niek K.C.B.R.

Contents

Foreword

During the past decade interest in quality management has greatly increased. Certification of quality systems on the basis of ISO norms and Quality Prizes have certainly contributed to this trend. Companies which have taken the initiative have seen that this forms the first step towards Total Quality Management (TQM). They have also seen that the application of statistics to these endeavours is absolutely essential. One of the central elements of TQM is Statistical Process Control, more commonly known as SPC.

This book describes the pitfalls and traps which businesses encounter when implementing and assuring SPC. We illustrate our approach from practical experience in various companies.

In essence, the management of processes amounts to the delegation of tasks, qualifications, and responsibilities wherever the processes take place. The integral approach is based on this fact and aims at achieving control and improvement. It is almost inevitable that the existing operations management will need drastic adjustment. Support from the executive management during implementation is therefore imperative.

Besides giving a general framework, we describe a plan of action for achieving statistically controlled (*i.e.*, predictable) processes. Special attention is given to the instruments which can be used for this purpose. Because the method is illustrated with many practical problems, application forms the central theme. Since an operations management based on SPC does not come to an end after one or two years, the book explains how the results which have been achieved can be assured and further improved.

The book is written for companies which are going to implement SPC, or which need a new impulse in order to get SPC properly off the ground. The following subjects are discussed:

- **Implementation of SPC.**
  SPC is placed within the correct framework. The approach required for implementation, the necessary changes to operations management and the role of the executive management are explained. The specific structures and activities and the practical examples form a convenient guideline for the implementation of SPC in companies.

- **Activity plan for achieving statistically controlled processes.**
  On the basis of an activity plan, a systematic procedure is described by which SPC can be introduced to the shop floor by the employees in question.

- **Statistical tools.**
  For the application of statistical methods in SPC it is important to bridge the gap between standard theory and practice. The descriptions provide insight into the inevitable consequences for the evaluation of the accuracy of measurements, the design of control charts and the analysis of process capability. By means of the examples and activity plan many practical situations can be tackled.

- **Consolidation and improvement of the results.**
  After a team has fulfilled the assignment to make the process step statistically controlled, the results must not only be retained but also improved. A structure is described which is in keeping with the introduction of SPC. Those aspects which are important for possible computerization are described. Further, a questionnaire for choosing the improvement team of the year is provided. Finally, we also provide a checklist with which a business can determine to what extent it has succeeded in the actual application of SPC.

This book has been based on our Dutch book entitled *Statistische procesbeheersing in bedrijf*. The Dutch book appeared in 1996 and has guided many companies in the Netherlands (and Belgium) with the implementation of SPC. There has been a strong demand for this book to be translated into English. We hope that we have succeeded in writing a book which can be used internationally. The typical Dutch situation has been transformed into a situation which is common for Western industries. However, the Dutch situation will sometimes be used as an example for a typical European country.

We would like to thank ASM Lithography at Veldhoven, Hollandse Signaalapparaten at Hengelo, and Philips Semiconductors at Stadskanaal for the opportunity to obtain so much experience with Statistical Process Control that we became able to write this book.

November 1998

RONALD DOES
KIT ROES
ALBERT TRIP

Introduction

The success of the application of SPC in Japan and the eagerness with which the United States subsequently rediscovered SPC made us realize that this approach was of essential importance for our companies. That our customers also urged us to implement SPC provided extra motivation.

The introduction of SPC requires a significant time, during which resources, a well devised approach, and a clear goal are essential. Making time and resources available did not prove to be the biggest problem. The manner in which SPC should be implemented was our greatest cause of concern. We succeeded in finding an approach which has been successfully implemented in our companies. This approach distinguishes itself by SPC being an operational system for the shop floor itself, and not a system for experts which, when it has finally been applied, has drifted far from the situation in which it really belongs. The results of SPC have not only become manifest in a reduction of waste, shorter throughput times and greater reliability of supply, but also in a better understanding of the processes and their variation. Furthermore, where both quality control and quality improvement are concerned, clear communication has been initiated between the various departments. By means of SPC we have succeeded in introducing an operational management of the continuous improvement of processes.

This book describes the manner in which we implemented SPC. The authors have succeeded in elucidating the various points of view. They make use of practical examples which have their origin in our companies. A very readable book has thereby come into being, and which we strongly recommend.

June 1998

P.N.R. KEEGSTRA
Director of Operations
Hollandse Signaalapparaten
Hengelo

M.L. VAN OORD
Operations Manager
Phillips Semiconductors
Stadskanaal

A.C. OUWERKERK
Vice President, Operations
ASM Lithography
Veldhoven

Chapter 1

SPC: A Historical Perspective

1·1 SPC Today

In this introduction we will take a look at the way in which SPC took off. To do so, we must go back to the 1920s, when mass production was introduced. This was a period during which people began to realize that in order to achieve a reduction in the number of defects it was not the products themselves that should be examined but the processes generated by these products.

For this, control of variation in the underlying processes is essential. In 1924 Dr. Walter A. Shewhart developed the control chart in order to realize process control at Western Electric, a company which had become specialized in the telephone market. The control chart is an outstanding instrument to facilitate process control. It is a chart indicating the time at which a quality characteristic is measured on the horizontal axis, and the value of the quality characteristic on the vertical axis. By charting a quality characteristic as a function of time we can judge whether or not a process is statistically under control. Figure 1.1 shows an example of a control chart.

Here, CL indicates the central line, UCL the Upper Control Limit, LCL the Lower Control Limit. These control limits are calculated from the data and are essentially different from so-called specifications. An extensive discussion of control charts is given in Chapters 6, 7 and 8.

SPC is, however, much more than the use of control charts. SPC is based on a management philosophy which drastically influences operations management. Owing to increasing competition on the world market, many companies have now switched to an operations management based on SPC. In some sectors production would now be impossible without SPC. The semiconductor industry is a good

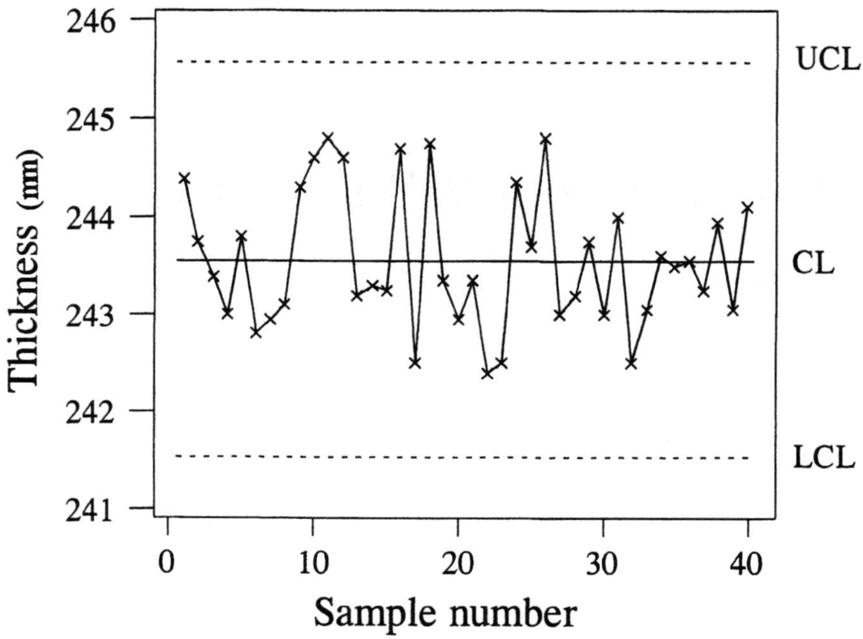

Figure 1.1: Example of a control chart with central line and control limits.

example of such a sector. It would be an illusion to assume that integrated circuits (ICs) could be mass produced without an operations management based on SPC.

In this book we present our experience with the implementation and assurance of SPC. As illustration we will use the experience gained in three companies which form a cross-section of (Dutch) industry. These three companies can be placed in the categories of mass-production, low volume production, and small series (or one off) production.

The first company (mass production) we present is Philips Semiconductors at Stadskanaal (part of Philips' Semiconductor Product Division) where more than 1200 persons are employed. A large variety of diodes are produced at Stadskanaal. Amongst other things, diodes are used as rectifiers and for protecting electronic apparatus against, *e.g.*, a lightning strike. A total of more than 3.5 billion diodes are manufactured annually, so that placing Philips Semiconductors in the mass production category is more than justified. SPC originated in this business category. Production is divided into six operational groups which, together with the R&D, Purchasing, Logistics and Marketing departments, form the primary production process. They are supported by the HRM, Facilities, Finance, Administration, Industrial Support departments, and various Laboratories.

Philips Stadskanaal can be characterized as an organisation which is constantly changing. In 1987 the Stadskanaal Integral Organisation Renewal was instigated, with the intention of altering the internal organisation drastically, and of defining product–market combinations. In 1990 the 'Vital Stadskanaal' programme was started. This revitalisation, within the framework of the worldwide restructuring of Philips (Operation Centurion), yielded a reorientation for turning a single factory into a business. This meant working on a customer oriented basis, improving results and developing the (order) portfolio. In 1994 a new programme ('Go for One') was started in which the main plans for expansion up to the year 2000 were outlined.

As far as quality is concerned, Philips Stadskanaal now has a reputation to keep up (expressed by being placed in the finals of the 1994 Dutch Quality Prize). Modern quality techniques have been applied since the end of the 1980s. The quality system was certified in 1991 as being in accordance with ISO 9001 and effectively working as such. Another important subject these days is environment. Philips Electronics realises this, and obliged all its sites to become ISO 14000 certified. In 1997 the Stadskanaal organization was one of the first within the company to achieve this result.

One of Philips Semiconductors Division's customers is Ford. This firm requires that its most important suppliers meet a number of conditions which exceed those stipulated in ISO 9001. For its suppliers Ford set up a quality programme called 'Total Quality Excellence' (TQE). Meanwhile, Ford, Chrysler and General Motors have now joined forces, and have replaced TQE by an even higher reaching programme, 'Quality System 9000' (QS-9000). In future, only those firms which meet the QS-9000 requirements will be allowed to supply the American car industry. At the end of 1991 the Management of Philips Semiconductors Division decided that the TQE status should be obtained. Naturally, Philips Stadskanaal also participated. In 1992 the Stadskanaal management decided that as a part of this goal SPC would be implemented in the way described in this book, which finally resulted in achieving the QS-9000 certificate at the end of 1997.

The second company we present is ASM Lithography at Veldhoven (low volume manufacturing). This firm develops, produces, sells, and provides service in the field of photo-lithographic projection systems known as 'wafer steppers', which are of essential importance for the production of advanced Integrated Circuits (ICs). These 'steppers' consist of parts with an extremely high grade of technological innovation, which has a positive effect not only on the local suppliers but also on medium sized international companies. ASM Lithography makes very high demands on its suppliers; their specifications are in nanometres. For the production of a wafer stepper about 90% of the parts are purchased, and of these 100% must conform to ASM Lithography's quality requirements. The company supplies wafer steppers to IC manufacturers in the US, Asia and Western Europe, and also offers its clients a complete support service package. About 2500 persons are employed by the firm.

The company can be grouped in the category of low volume manufacturing/fabrication. Four to seven wafer steppers are produced per week. Despite the low numbers, SPC can be applied successfully. Unfortunately, it is largely assumed that SPC works only when applied to large numbers.

At the end of 1991 ASM Lithography began with the implementation of 'Total Quality Management' (TQM). Various parts of the TQM programme fit in perfectly with what is necessary to implement SPC. A phased approach, awareness in management, a steering committee, inter-personal skills, thematic approach in teams, SPC methods and techniques, an improvement strategy and activities based on facts instead of on intuition and suggestion: all these are also ingredients needed to implement SPC. In 1993 the management decided to implement SPC in the way described in this book.

The third firm we present here is Hollandse Signaalapparaten at Hengelo (small series and one off fabrication). Hollandse Signaalapparaten, in short Signaal, is part of the French company Thomson-CSF. With a total turnover of 6 billion dollars and approximately 45,000 employees, the combination Signaal–Thomas is the largest in Europe and second in the world in the area of defence electronics. Signaal Hengelo is specialized in naval weapon control systems and radar systems; about 3000 persons are employed. The quality system is certified by Lloyd's as conforming with the ISO 9001 norm. Furthermore, Signaal also conforms to the specifications laid down by the Ministry of Defence (Allied Quality Assurance Publications, AQAP-110).

One of the production departments is the Small Unit Production (SUP). This department is specialized in the manufacture of customer-specific print panels, also known as Printed Circuit Boards (PCBs). A defence system often requires only one example of a particular type. On average, only 2.5 units of a particular type are manufactured, and each day a new print panel is designed at Signaal. Production within SUP is thus characterized by small series manufacture. However, since SPC focuses on processes and not on types, the number of products per process step is relatively large.

In 1994, under the terms of an order for the aircraft builder Lockheed Martin, Operations Management decided to implement SPC in the SUP in the way described in this book. The SPC certificate was obtained from Lockheed Martin in September 1995. Signaal introduced SPC integrally after 1995.

All companies mentioned above used the same approach. From our experience with other firms, *e.g.*, various branches of Philips both in the Netherlands and abroad, Douwe Egberts (part of Sara Lee) at Joure and Utrecht, Fokker Elmo at Woensdrecht, Fokker Aerostructures at Papendrecht and Schiphol-Oost, Heye-Glas Nederland at Moerdijk, Stork Digital Imaging at Boxmeer, we have learned that this approach is universally applicable. Obviously a custom made plan is needed for each firm, but the basic principles given in this book can serve as a guideline.

Despite this fact it is striking that SPC has not yet become truly established within industry. Excuses which are often heard are:

- Our company is too small for SPC.
- Our numbers are not large enough.
- My personnel are insufficiently educated to use the statistical tools.
- My clients are only interested in costs.
- We tried, but it did not catch on.
- We are busy enough as it is!

In this book we shall show all these excuses to be nonsense. Correct implementation of SPC will yield a company the following results:

- A decrease in waste.
- A better distribution of tasks, competence and responsibility within the company.

□ Improved relationships with customers and suppliers.

□ A basis of improvement.

□ Survival!

We end the section with the last point (survival). If a company waits too long before implementing SPC the penalty could be extremely high. Compare it with the preventative approach of changing the oil in a motor car, costing a few dollars. This can be deferred at the risk of a defective motor. Cost: a few thousand dollars or, worse still, having to walk or take the train in future!

1·2 SPC in the Period 1924–1945

As mentioned in the previous section, the development of SPC in 1924 started with the introduction of the control chart for process control. The first classic book on SPC was published in 1931. This book, entitled 'Economic control of quality of manufactured product' by Walter A. Shewhart, can also be seen as the beginning of industrial statistics. Shewhart concentrated on production processes. Shewhart worked for 18 months at Western Electric in Chicago. Part of the product range of Western Electric was 'telephone parts'. The business was ambitious enough to advertise with the slogan 'as alike as two telephones'. However, they found that the harder they tried to realize this goal, the worse the results were. The more effort put into limiting the variation, the larger it turned out to be. Shewhart, by then working for Bell Laboratories in New York, was commissioned to develop a form 'which might be modified from time to time, in order to give at a glance the greatest amount of accurate information'. After thorough analysis it became clear to him that the problems at Western Electric could be differentiated into the following two categories:

□ that an error, complaint, mistake or accident was treated as it if had a special cause, while in fact it was caused by random variation owed to common causes.

□ that the error was treated as if it was owed to common causes, while in fact it was owed to a special cause.

It is essential to distinguish clearly between common causes and special causes. Shewhart discovered that not making this distinction was precisely the main cause of the problems experienced by Western Electric while trying to attain a decrease in variation. Because this is so important, we will elucidate the distinction in two ways, on the one hand by means of an illustrative example, on the other hand with a more fundamental explanation.

In his inaugural address entitled 'Industrial statistics guarantees quality' (Does, 1992), Professor Ronald J.M.M. Does asked his audience to "suppose that right now I was able to measure your attention paid to this lecture. Your attention is influenced by factors such as the temperature in the auditorium, the lighting, how comfortable the chairs are, my style of delivery and this topic. In short, factors which are the same for everybody. There are, however, factors which may influence the attention of an individual, such as lack of sleep, family problems and health. In short, factors which are incidental. If the lack of attention is primarily owed to common causes, then I and those responsible for the design of the auditorium must take action to ensure better attention in future lectures. If the lack of attention is primarily owed to special causes, then the audience should take action."

This example is given with the intention of showing how important it is to know whether the variation is the result of common causes or of special causes, since the action to be taken depends largely upon this knowledge. Shewhart laid the foundation for this distinction. The definitions we use are (see also Nolan and Provost, 1990):

Common causes Causes which are inherent to the process hour by hour, day by day, and which influence everything and everybody working in the process.

Special causes Causes which are not continuously present in the process or which do not influence everybody, but which arise from specific circumstances.

In this context we use the following description for the concept of 'process':

'A *process* is a collection of mutually related resources and activities which transforms input into output'.

The result can be either a product or a service. For each process the performance and the quality characteristics should be identified and measured.

Examples of quality characteristics in manufacturing processes are: thickness of a layer of tin, width of copper tracks on a PCB, diameter of a glass bead, time of fall of the weight on a elevator, concentration of a chemical bath. Examples in service processes are: time needed to handle a client's query, amount of erroneous billing, waiting time at a counter, number of erroneous reservations.

We see that in all examples the results of the quality characteristic can change with time, and this is where 'variation' enters the discussion. The core of SPC is that we are only able to deal properly with variation if we understand the difference between both types of variation (variation resulting from common causes and variation resulting from special causes).

To illustrate the point, consider a process the purpose of which is to grind wafers (slices of silicon) to a given thickness. The grinding process has been analyzed during a given period (a random wafer is measured in each grinding batch). In Figure 1.2 measurements of the thickness of a number of grinding batches taken during this period are plotted.

The question is whether the last measurement, at which the arrow points, gives rise to intervention. In practice we should ask ourselves this question each time. Measurements are taken in order to make decisions on the basis of the results. If we take measurements for their own sake only, we may as well cancel them. Valuable manufacturing time will then be saved. This is, however, not true for this grinding process. Each measurement gives occasion to ask: intervene, or not? Looking at the final measurement, it is striking that this does not deviate greatly from the other measurements.

Besides the measurements carried out by himself, the operator also has access to the specifications, namely, the minimum requirements which must be met. The specification limits are given in Figure 1.3 (LSL represents the 'Lower Specification

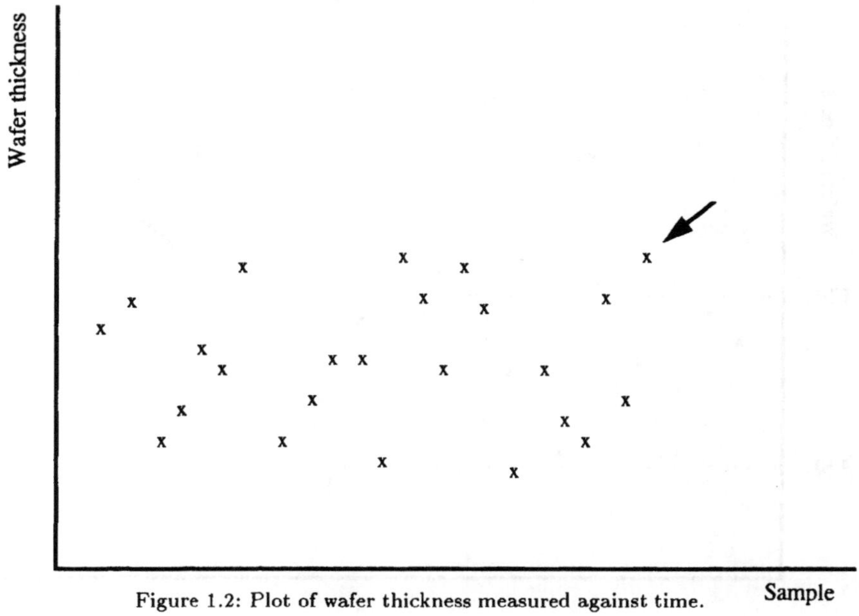

Figure 1.2: Plot of wafer thickness measured against time.

Limit', USL the 'Upper Specification Limit'). In our example, the specifications are LSL = 235 micrometres and USL = 255 micrometres.

From Figure 1.3, it can be seen that the wafer is too thick. Apart from possible repair, we must view this wafer as waste. We should now ask ourselves whether we need to adjust the process since we have produced waste, or whether we should just go ahead with the next grinding batch. To make the problem even more complex we will consider a second, completely different, process.

The second case concerns the application of copper tracks on a PCB. To accomplish this task the pattern is projected onto a sensitized PCB using a film of the track layout; the system is then exposed and the PCB 'negatives' developed, after which subsequent etching and washing steps in the process ensure that the correct copper track remains. In this second process the concentration of sodium carbonate in a developer is registered as a quality characteristic. For these measurements the specifications are: LSL = 0.8% weight and USL = 1.0% weight. Here again, measurements were carried out during a given period. The results are given in Figure 1.4. In Figure 1.5 the specifications are added to the same measurements.

Figure 1.4 shows that the final measurement deviates greatly from the previous measurements. However, as can be seen in Figure 1.5, it turns out that this final measurement actually falls within the specifications.

Situations 1 (the grinding process) and 2 (the developing process) illustrate the difference between variation owed to common causes and variation owed to special causes. Although there is absolutely no waste or suchlike in situation 2, it is striking that the final measurement deviates greatly from the other measurements. This is precisely the opposite in situation 1: waste is indeed produced because the measurement falls outside the specifications, but in itself the last measurement

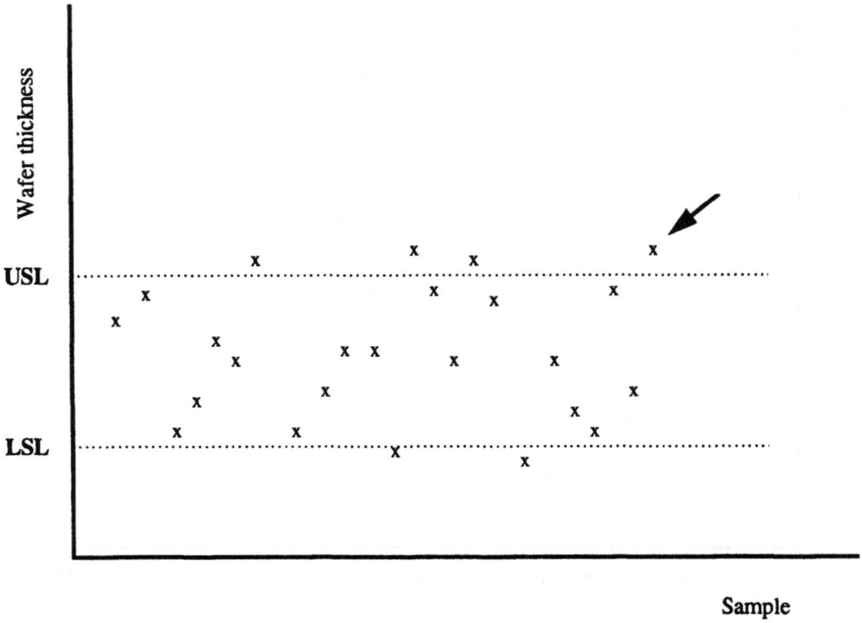

Figure 1.3: Plot of the wafer thickness. The specification limits are also shown.

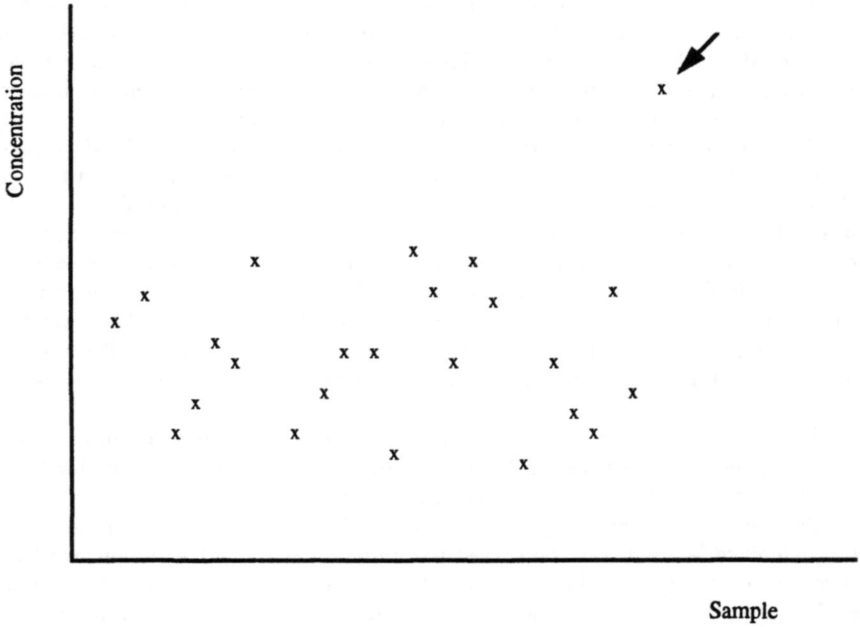

Figure 1.4: The concentration of sodium carbonate plotted against time.

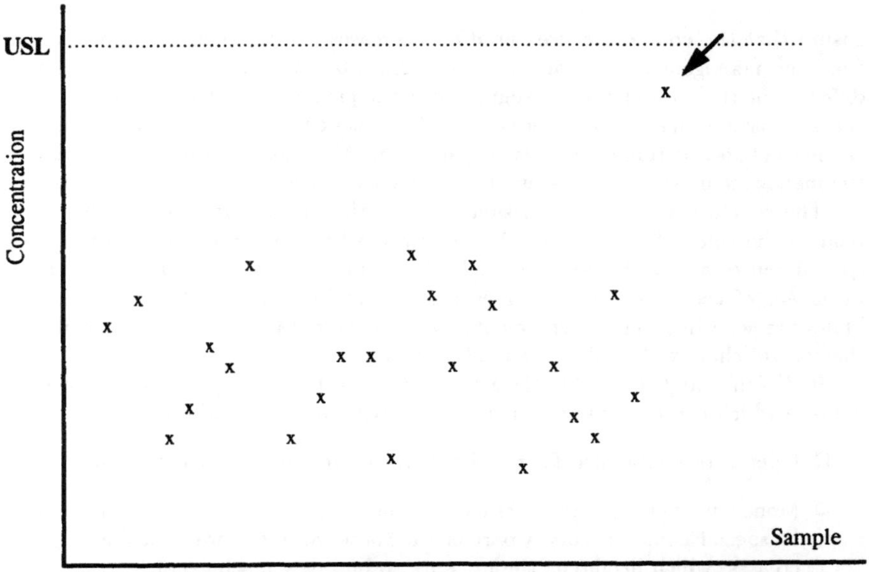

Figure 1.5: Plot of the concentration of sodium carbonate (including specifications).

does not deviate greatly from the previous measurements. Figures 1.3 and 1.5, in which the specification limits are given, cause the operator's attention to be diverted from what is really necessary: to distinguish between the two types of cause.

Specifications establish the acceptable variation, independently of the variation found in the process itself. Specifications can also be called the 'Demands of the Customer', where 'customer' can also refer to the next step in the process and not *a priori* to the end user. But as the given examples show, the demands of the customer does not necessarily fit the process. The 'Voice of the Process' is determined by plotting the quality characteristic against time. In this way we obtain insight into the possibilities of the process, or, in other words, the variation which belongs to the process. This is often called common cause variation. Figures 1.2 and 1.4 give an example of common cause variation. In Figure 1.2 we can see that the final measurement fits in with the voice of the process. By intervening in this situation we would introduce extra variation which does not belong to the process. The result is: extra waste. This is obviously not what we intend to achieve by intervening, but it gives an explanation for the problems at Western Electric in the 1920s. In pursuit of greater uniformity they intervened, while in fact the variation was inherent to the process. The result was even more waste. From Figure 1.4 we can see that in the second situation the process was disturbed by some special cause. Direct intervention is necessary to locate the cause and to let the process continue as before.

In short, in the case of variation owed to special causes direct intervention is necessary, while in the case of variation owed to common causes intervention is counterproductive. Note that in the first situation (the grinding process) we still have a problem. The variation inherent to the process is so great that waste is produced. The process is simply unable to comply with the demands of the client. To solve this dilemma an improvement in the process must be sought in order to ensure that the common cause variation decreases. This usually requires action from the management who are, after all, the process' owners. The situation is different in the case of the photographic developing process (situation 2). Here action must be taken by the operator. It is his task to ensure that the process is carried out in accordance with its capabilities. Also, some action on the part of the management will be necessary to prevent the special cause from recurring.

The conclusion of this discussion about variation is that attention must be paid to the voice of the process. Differentiation between variation as a result of special causes and variation as a result of common causes can easily be attained by means of the control chart. This means that in Figures 1.2 and 1.4 control limits are set which have been calculated from the data. In Chapters 6, 7 and 8, the control chart will be discussed extensively.

In Nolan and Provost (1990) a number of examples is given of the type of damage which can be caused by incorrect interpretation of variation:

- Operators are blamed for problems which are outside their influence.
- Money is spent on unnecessary new machines (when the variation owed to special causes becomes part of the total variation, waste ultimately increases, which heightens the pressure to buy new machinery).
- Time is wasted looking for the cause of a signal, while in fact there is nothing wrong.
- The variation is attributed to specious reasons, to which it has no real relation.
- Action is taken, where it would be better to do nothing.

It is thus of great importance that every organisation understands how to cope with variation. This is especially of crucial importance for the management, given the damage which can arise if this knowledge is not at its disposal. Not understanding the distinction between both types of variation can be the difference between success and failure.

In the initial period of SPC (about the first twenty years after 1924), it was seen as a statistical technique which could be applied well to manufacturing processes. During the Second World War the control chart was implemented on a large scale in the United States. One of the most prominent quality gurus from Japan, Ishikawa, stated that it was even possible to speculate that the Second World War was won by the Allies because of the impact SPC had on the quality of their weapons (Ishikawa, 1985).

Although statistical methods had more than proved their worth in the weapons industry during the Second World War, they were not applied to the consumer industry in the period following the war.

The only country able to manufacture goods at that time was the USA. In general, it was thought that the more one produced the more it was possible to sell. Quality was made subordinate to quantity. In Western Europe the recovering industry also did not realize that SPC was the perfect method to safeguard and improve quality.

Things were very different in Japan. After the war, Dr. W Edwards Deming, a disciple of Shewhart, was invited by General MacArthur to act as a consultant for Japanese industry as part of the American help programme. In the following paragraph we will describe the approach used by Deming.

1·3 SPC after 1945: Out of the Crisis

Deming, already mentioned in the previous paragraph, is the father of the third wave of the industrial revolution, and is celebrated by the Japanese as the architect of their phenomenal success (although Banks (1993) disqualifies this claim somewhat). Because of his special merit we give a short life history here (for an extensive life history, see Neave, 1990).

The American William Edwards Deming was born on 14 October 1900. He began his career as a university lecturer after studying mathematics and physics. He gradually became more interested in applied statistics, which brought him into contact with Shewhart. This interest stood him in good stead when he accepted a position at the Ministry of Agriculture in 1928. After ten years he transferred to the American Bureau of the Census. In both positions he applied Shewhart's theory in an environment different from the traditional manufacturing setting. In the disastrous year of the attack on Pearl Harbour (1941) Deming, together with a number of colleagues, managed to convince the weapons industry to make use of Shewhart's approach. Deming was convinced that the war could be won if the weapons were of excellent quality. In his book his Japanese colleague Ishikawa (1985) frankly admits that Deming was right.

Deming's broad background and his reputation within the government led to his being asked to participate in the post-war reconstruction of Japan. This began his career as a 'Consultant in Statistical Studies'. After the war the Union of Japanese Scientists and Engineers (JUSE) regularly invited Deming to give seminars on SPC. He started with these seminars in 1950. Since Deming had experienced that his philosophy was not adopted by executive managers in the United States, he now tried to avoid that recurring. The chairman of the JUSE, Ishikawa, was therefore requested to call a meeting with members of the Kei-dan-ren, the Club of Japanese 'Captains of Industry'. The first meeting took place on 13 July 1950 with twenty one Captains of Industry. During the summer of that year Deming spoke to about 100 Japanese senior managers. In 1951 about 400 Japanese managers followed. From the start, appreciation of this contribution to the economic development of Japan was great, and was expressed in *e.g.*, the instigation of the 'Deming Application Prize'. Since 1951 this prize has been awarded annually to businesses and persons who have an excellent performance in the field of quality. In 1960 Deming received from the Japanese Emperor himself the highest honour which can be awarded to a foreigner, the 'Second Order Medal of the Sacred Treasure'.

From 1950 onwards Deming did not restrict himself to statistical methods. He developed a much broader point of view in the field of quality, a point of view which is known today as 'Total Quality Management' (TQM), referred to in this book as SPC. Deming's opinions are based on the fundamental ideas of Shewhart. Like no other, Deming understood the difference between variation owed to special causes and variation owed to common causes. Special causes are often found in the area which can be influenced by operational personnel, and are either a systematic error or an unexpected disturbance of the process. By eliminating special causes a process is created which can be called 'statistically under control'. What remains is variation owed to common causes. This type of error (also called system errors) is directly within the area which can be influenced by the management. From many studies in the 1950s and 1960s Dr. Joseph Juran calculated that for work on the shop floor the ratio between variation which can be influenced by management (common causes) and variation which can be influenced by personnel (special causes) was about 80:20 (Juran, 1989). Deming, always more radical in his opinions, corrected this ratio to 94:6.

Deming (1986) formulated fourteen points aimed at creating a better environment for the process of improvement. These points can serve as a guideline for management when carrying out their tasks. They can be universally applied, in both small and large businesses, in the provision of services and in factories. Given the importance of this approach, the fourteen points are elucidated here:

(1) Be constantly aware of your goals for the improvement of product, process or service in order to stand up to competition, facilitate continuation of the business and ensure employment.

(2) Adopt the new philosophy. We live in a new economic era created by Japan. The management of Western businesses must ensure it can cope with the new challenges, and should be aware of its responsibilities and show leadership when making the necessary changes.

(3) Stop being dependent on inspection to achieve quality. Aim to make inspection superfluous by making quality an integral part of the product or service in the first instance.

(4) Do not judge suppliers by their price tag only. Cooperate with suppliers in long term relationships based on loyalty and mutual trust. Only then will it be possible to learn together. Cooperation is an absolute necessity.

(5) Improve the manufacturing system and the service continuously for better quality and to diminish the costs.

(6) Institute training on the work floor. Redefine training programmes. The management requires training to learn all about the business, from incoming raw materials to the customer. The distinction between variation owed to common causes and variation owed to special causes must be fully understood. For operational personnel, training is impossible without the use of work methods. These employees must be trained so that they are able to keep the processes statistically under control and to eliminate possible special causes.

(7) Introduce leadership with the aim of helping employees and machines to do things better. The task of the manager is to take the lead in helping employees, coaching them and letting them use the best methods known. Evaluation, in the sense of reward and punishment for results, is outdated. Managers have a new task to master.

(8) Ensure that fear is eliminated, since fearful people will not give true and trustworthy data, and SPC will not work without reliable data. Employees

must feel free to ask questions, to ventilate ideas, to ask for explanations about the work and work methods and to report on factors which form an impediment for good products or good services.

(9) Break down the barriers between departments. If each department is separately optimized, the whole will remain disjointed. Everything must be optimized, the whole organization, together with the suppliers and customers, as one coherent system.

(10) Stop using slogans, exhortations and target figures which tell personnel achieve 'zero defects' or new productivity levels. These encouragements are counterproductive, because the majority of reasons for bad quality and low productivity originate from the system and therefore lie outside the field of influence of the employee.

(11) Get rid of work norms which enforce target figures. Switch over to leadership (see point 7). Assignments such as 'produce so much a day' or 'you have three hours for this job' must be replaced by true leadership. Managers must also refrain from statements such as: 'I want these results; you figure out how to achieve them.' Managers must cooperate with the personnel to develop increasingly better methods.

(12) Remove obstacles which deny employees pride in their professional skills. The responsibility of management must be focused on quality instead of quantity. This also means dispensing with annual evaluation of employees. The performance evaluation system is a source of distrust and creates unnecessary dividing lines between employees and departments. Performance evaluations are win/lose concepts. Aim at win/win situations.

(13) Introduce an effective education, training and self-improvement programme. There is a marked difference between this point and point 6, which relates to training in order to enable personnel to carry out their work well. Here, we refer to education which generates ideas and improvement of everyone's work.

(14) Let everyone in the business be involved with implementation of the changes. They affect everyone's job.

If these fourteen points are used as a guideline lasting improvements will be achieved. The question arises of whether all points must be realized simultaneously. Deming developed these points in the course of about twenty years. They are meant to remove obstacles which repeatedly occur during the improvement process and they can be realized whenever a need arises. Ultimately they must all be realized.

One of the leading advocates and propagators of Deming's management philosophy is Dr. Brian L. Joiner (Joiner, 1994). He uses the so called Joiner triangle (which he devised together with his wife) to emphasize the central elements of his outlook. Our version of the triangle is given in Figure 1.6.

In our terms the three angles of this figure are Quality, Variation and Together. In a well adjusted system these three are not separate, but inseparably connected to each other. The heartbeat in this triangle is SPC. Neave (1990) indicated that

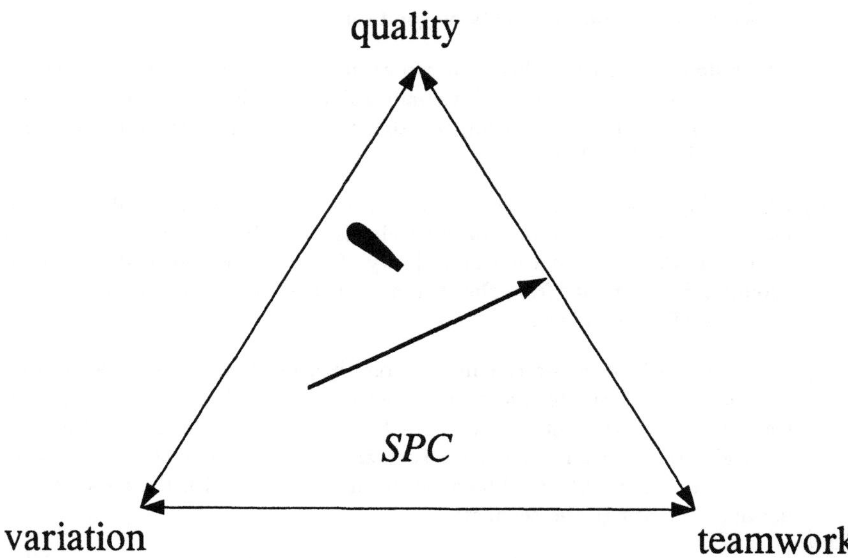

Figure 1.6: The Joiner triangle.

this triangle can also be used to cluster Deming's philosophy. The obsession with quality is expressed in the first six points and points 13 and 14. How variation can be dealt with is conveyed in points 10 to 12. The remaining points, 7 to 9, are about working together, operating as one large team.

Under the motto "My people are destroyed for lack of knowledge" (Hosea 4:6) Deming (1986) described some more fatal diseases which obstruct the alteration process of his management theory. Amongst others he names:

- Lack of a constant goal to plan products and services for which a market exists, which will keep the business going and create employment.
- Emphasis on short term profit and short term reasoning.
- Evaluation of performance, annual evaluation of employees.
- Supervision on the basis of figures only, so that factors which are more difficult to measure, such as work satisfaction of employees and motivation, are not included.

Apart from the fatal diseases, Deming also names some other obstacles which companies put forward in order not to implement SPC or to argue that SPC does not work or is not necessary. A few salient points are:

- the hope for instant pudding;

- □ requiring examples first;
- □ inadequate education in statistical methods in industry;
- □ use of the military standard 105D and other sampling plans;
- □ our Quality Assurance Department takes care of all quality problems;
- □ only shop floor personnel have problems;
- □ we have installed quality management;
- □ the computer takes care of quality;
- □ we need only conform to specifications;
- □ the misconception about zero defects;
- □ the person who can help us must know all about our company.

It was only at the end of the 1980s that people in the West began to realize how important Deming's approach is for satisfying customers, employees, and shareholders. The American company Nashua, under William E. Conway, was at that time the first company where Deming worked as a consultant in the area of SPC in the broad sense. He was already seventy eight years old, an age at which most people begin to take their ease. The rest of American business was rudely awakened by the NBC documentary 'If Japan can, why cannot we?' on 24th June, 1980. Given his age, Deming was chiefly oriented towards large corporations during this period, the reason being that large corporations have many suppliers, so that in this way his message would be spread more quickly (Boardman, 1994). A few of these large corporations were Ford, General Motors, Hewlett Packard, Dow and Procter and Gamble. After the NBC documentary, Deming again began to proclaim his ideas to a wide public. His four day seminars were especially popular. In this way, an estimated 10,000 people were introduced to his approach.

He did not lack awards. Besides the already mentioned Japanese honours, Deming received the National Medal for Technology from President Reagan in 1987. In addition he was awarded 20 honorary doctorates, the 'Shewhart medal' and Honorary Membership of the American Society for Quality. Deming continued proclaiming his message actively up to his death on 20 December 1993.

In the Netherlands interest in SPC arrived later than in the United States. The executive manager of Volvo Cars, A. Deleye, was so deeply affected by an article in one of the Dutch papers on 15th July, 1981, that he immediately went to Nashua to see Deming and Conway. In 1983 the MANS Foundation was set up, which went on to propagate the ideas of Deming in The Netherlands, adjusted, where necessary, to The Netherlands and Dutch relations. It gave a new impulse to quality thinking in the Netherlands.

We can state now that the necessity of taking quality seriously is generally accepted in the Netherlands. A few consultancy firms have specialized in this area. The universities are also taking initiatives for active participation in this market. The University of Amsterdam has even founded the Institute for Business and Industrial Statistics (IBIS UvA BV).

Chapter 2

SPC as Part of Quality Policy

2·1 Quality, Dealing with Variation and Working Together

Joiner's triangle (see Figure 1.6) contains the three elements of Deming's management theory: quality, variation and teamwork. In this section we will take a closer look at these three parts, using ISO 9000 as a basis. These norms have been valid since 1987 (cf. Bureau of Business Practice, 1992) and a new version was published in 1994. As far as responsibility for quality policy is concerned, the requirements are very specific: this lies with the highest management level, whose task it is to create a climate for continuous quality improvement. Managers radiate leadership and commitment through their own deeds, perseverance, and deployment of resources. They direct quality improvement by

- Conveying the purpose and the goals;
- Improvement of their own work processes;
- Promoting a climate of open communication, teamwork and respect for the individual;
- Enabling and authorizing all within the organization to improve their work processes.

This definition from ISO 9004-4 (Guidelines for Quality Improvement) is perfectly in keeping with what emerged from the discussion on Deming's management theory.

We can also fall back on ISO 9000 for a definition of quality. Our definition here is more or less identical: 'Quality is the degree to which the totality of characteristics of a product, process or service conform to the requirements resulting from their utility purpose'.

We are thus discussing a relative concept: 'quality is the degree to which'. To make this concept practically manageable, we need to translate it into specific, quantifiable variables: to what extent does the product, process or service conform to which requirements? In such a translation the concept 'variation' again emerges, since quality varies for products and processes as well as for services. Products differ, although they may seem identical: environmental conditions can vary, wear may occur, *etc.*. In short, variation is omnipresent, and plays a role in all we do. So reasoning about quality is inseparably coupled to reasoning about variation. The language of variation is best understood by making use of statistics and (this) explains the free use always made of statistical methods and techniques within quality management (see Montgomery (1996), for the role of statistics in sample inspection, process control and in the design of products and processes). Dealing with variation forms the link between statistics and quality management. The minimum requirements with which a quality characteristic must comply are set down in the specifications (section 1.2). In traditional quality management, people adhered to the principle that everything was all right as long as the values of the quality characteristics were within the specifications, and that things were wrong when these values were outside the specifications. It is the merit of Genichi Taguchi that we now use a better specific definition of quality.

Wheeler (1991) even calls Taguchi's approach a revolution in quality management. Taguchi (1986) indicated that the traditional way of viewing the cost of variability (based on the 200 year old definition of quality as 'conforming to the specifications') was utter nonsense. In the traditional view, if a quality characteristic has a value which conforms to the specification, the loss of quality is zero, while the costs suddenly rise markedly because of down time, re-processing, tracing the cause of the defect, repairing the defect and the time and energy needed to write reports and memoranda on the problem.

In September 1960 Taguchi defined *quality* as:

'conformance to a specified target value',

and *quality improvement* as:

'reduction of variation around the target value'.

His quadratic loss function (Figure 2.1) gives a more realistic picture of the cost of variability than the traditional view, and is now generally accepted.

The consequence of reasoning in terms of variation is that variation control, and thus statistics, play a much larger role than is usually recognized. Improvements are achieved by reducing variation. Dealing with variation involves more than the application of statistical methods and techniques. It especially requires insight into the mode of thought which forms the basis of these methods, and this is precisely what SPC advocates.

Dealing with variation must be carried out in a structured way. We choose a scientific method which has more than proved its worth in practice. To initiate an accumulative learning process within an organization, the well known *Plan, Do, Check, Act, Plan, etc.*, (PDCA) cycle, also known as the Deming cycle or the Deming wheel, is an excellent strategy (see Figure 2.2).

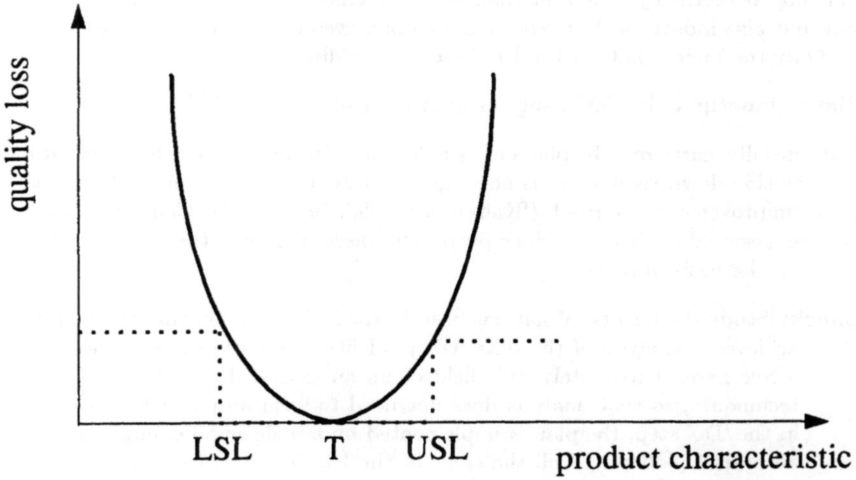

Figure 2.1: The quality loss function according to Taguchi (solid line) and the traditional loss function (dotted line).

Deming himself called it the Shewhart cycle because he felt that he had learned this principle from Shewhart. This cycle is discussed in detail in Joiner (1994). Joiner argues that the Deming cycle forms the mainstay of quick learning. The basic principles are so simple that anyone can learn them in five minutes. At the same time, Joiner added that, after 10 years experience, he still hoped to really understand the cycle one day!

The cycle is a scientific method which should be used everywhere within an organisation. The management plays a key role, because it defines the course to be taken in the pursuit of improvement of the organization. For both management

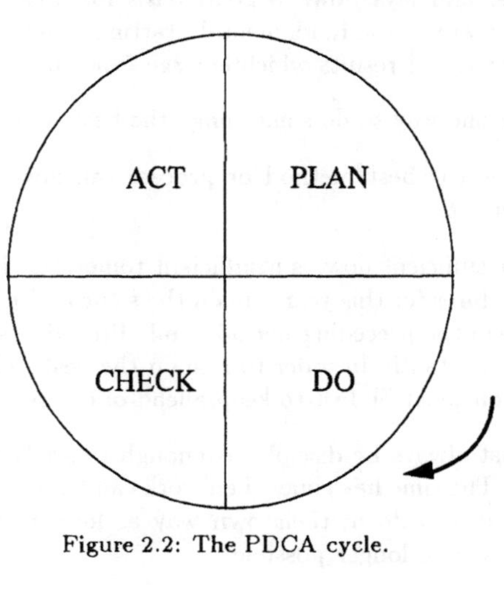

Figure 2.2: The PDCA cycle.

and shop floor the cycle is a method not only enabling today's work to be done well, but also indicating how work can be done even better in the future.

Only the main points of the PDCA-cycle are discussed here:

Plan: Draw up a plan for change or improvement.

Do: Initially, carry out the plan on a small scale. Turning the whole organization upside down right away is not a good strategy. This principle fits into the 'improvement' approach ('Kaizen principle', Imai, 1986). Collection of data is essential in this step, since purposeful discussions can then be carried out on the basis of facts.

Check: Study the results. What has been learned? Determine what the plan has achieved. Analysis of the data will yield firm conclusions. Statistics plays a role here. Fortunately, this field offers an extended arsenal of graphical techniques, so that analysis does not need to form an obstacle. Note that in the 'Do' step, the plan is implemented on a scale small enough to retain an overall picture. Of all the steps in the PDCA cycle, the 'Check' step is the most important. It is the catalyst for quick learning. Without a formal check, improvement is almost impossible.

Act: Intervene! It is important that we do something with the conclusions. In 'Check', we determined what we had learned. We now have the following choices: implement the plan, change the plan, or reject the plan. Often the cycle will be carried out again, if possible under other circumstances and conditions, to examine in how far the results are reliable and constant, and to see whether the tested approach also works elsewhere. Implementation of a change or improvement should be carried out carefully, but in practice this is often left undone. Retaining the newly obtained know how safeguards an organisation against making the same mistakes again and again.

Application of the PDCA cycle is the difference between 10 years' experience and the same experience repeated ten times in one year. A business must learn from its experience, and lays down a good basis for doing so by applying the PDCA cycle. There are a few fundamental starting points for establishing and integrating the educational results which emerge from the cycle:

- There is only one way to do something: the best way.
- What today is the best method or process can always be improved. No method is perfect.
- That which is sufficient now, is insufficient tomorrow. If progress is stopped and the procedures for this year remain the same as for last year, the pace of improvement in the preceding period is nil. Procedures and processes must be improved constantly in order to remain the best. The pace of change in an organization must be fast to keep ahead of or up with the competition.
- Everyone must always be disciplined enough to apply the best methods of the moment. The time has gone when work can be seen as art, as something which everyone can do in their own way as long as the desired result is achieved. This is no longer possible!

- □ Everyone must search actively for opportunities for improvement, since only then can the pace of development remain high.

The elements of Plan, Do, Check, Act are present in many organizations. However, the activities are usually spread over various departments. This creates imbalance because one department draws up a plan whilst another department is responsible for carrying it out, a third department is responsible for the measurements, *etc.*. Cyclic completion of the method is thus carried out in a way far from ideal. Only when everyone within the organization is involved in all four elements of the cycle will they feel committed to improving their activities. Again, we refer to Deming's 14 points, because they create a climate within an organization whereby the above can be realized.

In this section we have specifically explained what we mean by quality and dealing with variation with the help of a scientific method. We conclude by explaining the last element of Joiner's triangle, 'Teamwork'. An example of 'working together' is a survival trek. The usually primitive conditions with little luxury force the participants to work together, since otherwise the assignment will not be achieved. Running a business is, in many ways, similar to a survival trek. It means trusting our employees and treating them with respect. The role of the management is not so much motivating the employees as removing causes, circumstances, *etc.*which lead to de-motivation. We must assume that the employee is always prepared to do his best, and that the role of the management is to take away any barriers which obstruct this. Working together as a team is essential hereby. The Law of large numbers is valid here: group reasoning is usually better, less variable and more precise than individual reasoning.

The importance of teams within quality programmes is generally supported. Improvement teams increase the effectiveness of a business because all employees are involved in the improvement of professional performance. A few comments can be made here:

- □ Setting up improvement teams always meets with resistance. This is no reason to decide not to go ahead; however, it is a fact which must be taken into account.

- □ The management leaves the initiative to the personnel and thereby gives up a bit of responsibility. There is nothing wrong with this, but is can be taken as an encroachment on their own position.

- □ The aim is not only to have an eye on the financial results, but also to view the way in which the team functions. Cooperation of employees within a business is a factor which must not be underestimated.

- □ Security and stability in a business are dependent on rules, procedures and systems which have often been in force for years. They are, of course, necessary but they can also work as an impediment. If we wish to improve we must be prepared to ask questions about why we do things as we do and why we reason as we do. Statements such as 'this is the right way' and 'we always do it like this' do not mean that things must always remain so.

- □ Deming (1986) gave his chapter on the principles for change in Western management the motto 'How poor are they that have not patience'. This motto is just as valid for improvement teams. Give them time to remove bottlenecks.

- The management must ensure that the initiative and enthusiasm of the teams is preserved as far as possible. To achieve this it must make time and resources available and show interest in the approach.

Teams play a central role in the methodology presented in this book, by which a specific interpretation of how to control processes can be obtained. In Chapters 4 and 5 we will give a detailed description of how these teams operate and which resources they use.

2·2 An Operations Management Based on SPC

According to In 't Veld (1988) there are several reasons why a manager does not readily choose a new operations management, even when it is recommended as being the best.

Firstly, a manager never has enough time. Following a one day introduction to a new method is even asking too much. Reading a fat book on the subject does not fit into his daily routine. A manager usually runs his business on the basis of experience.

In the second place, and perhaps most important reason, it is that there is a gap between theory and practice. Today's manager wrestles with problems which are a result of the increasing complexity and changing attitude of both the individual and society. It takes quite some time before a theory is manageable in practice. A manager will be wary of using a new theory right away. It will first have to be demonstrated that the theory really works.

We have taken these points into account as much as possible in this book, by:

- Clearly, and in relatively few pages, indicating what we mean by SPC (Chapters 1, 2 and 3 are especially written for managers);
- Giving the reasons which companies themselves indicate for implementing SPC (see Section 3.1);
- Giving a phased approach for the implementation of SPC (see Section 3.2);
- Indicating how a phased plan can be carried out under the guidance of an SPC steering committee (see Section 3.3);
- Describing a complete activity plan (Chapters 4 and 5) which can be used as a guideline for the implementation of SPC by Process Action Teams (PATs), thus diminishing the gap between theory and practice;
- Singling out the statistics from the rest of the story and describing them separately from a practical point of view (Chapters 6 to 9). Spectacular results can be achieved with this set of instruments;
- Describing the assurance and consolidation, computerization, self-evaluation and improvement (Chapters 10 to 12).

What does an operations management based on SPC entail? Hoerl (1995) writes that SPC implies that:

- All the work is realized in a system of mutually connected processes, each having a customer/supplier relationship;
- All processes show variation;
- Sources of variation can be basically divided into variation owed to common causes and variation due to special causes;
- Understanding the origin of both sources of variation is the key to reducing variation, and
- Reducing variation is, in its turn, the key to quality improvement, productivity and profitability.

This train of thought it visualized in Figure 2.3.

In practice this means that SPC (Hare, Hoerl, Hromi and Snee, 1995):

1 Puts strong emphasis on the processes instead of on the resulting products;

2 Pays a lot of attention to the definition, control and reduction of variation;

3 Deals with problems on the basis of data.

This way of working is generally called 'statistical thinking' (Snee, 1990). Implementation of SPC in the broadest sense of the term therefore automatically leads to statistical thinking. SPC is thus much more than the application of statistical methods and techniques. The degree of acceptance is defined by actual and vigorous support of the management. The Joiner triangle—obsession with quality, dealing with variation, and working together—forms the basis for the approach. Implementation of SPC should be phased. The role of the management in this process is essential.

For SPC to be successful, the management must ensure that it reasons in terms of processes, that it understands how to deal with variation and that data is used as the basis for decisions.

Managers should have a good understanding of how to deal with variation. It is no coincidence that the pioneers (Shewhart and Deming, to name the most important) classify themselves as statisticians in the field of total quality management. Variation, special causes and common causes, SPC, process improvement by experiments, all these are concepts and areas where statistics play an essential role.

Managers who wish to control the way in which variation is dealt with need to be able to make a clear distinction between variation owed to special causes and variation owed to common causes. They then should obtain practical experience in their own field of work, which should not pose a problem seeing they receive a surfeit of data each day. Managers who have been trained in the basic principles of how to deal with variation should be able to plot this data against time and to determine the limits within which the data usually varies. In this way the management discovers what it really means to have processes with variation.

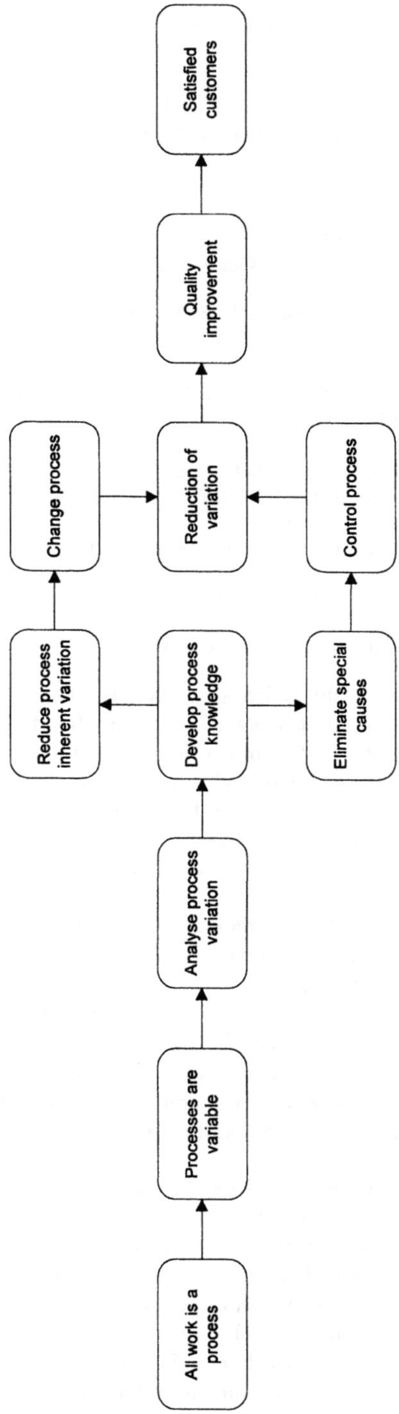

Figure 2.3: Statistical thinking in business.

The guiding role of management in the implementation of SPC is so important that it cannot be emphasized too much. It would be an illusion to assume that SPC could be implemented without the enthusiastic support of the management. For Deming this was the reason for formulating his 14 points. The majority of problems (94% according to Deming) are caused by factors outside the field of influence of operational personnel. Intervention by the management is then urgently needed to achieve a breakthrough.

Snee (1990) developed a model for total quality in which he describes the three fundamental levels of activity in an organization, see Figure 2.4.

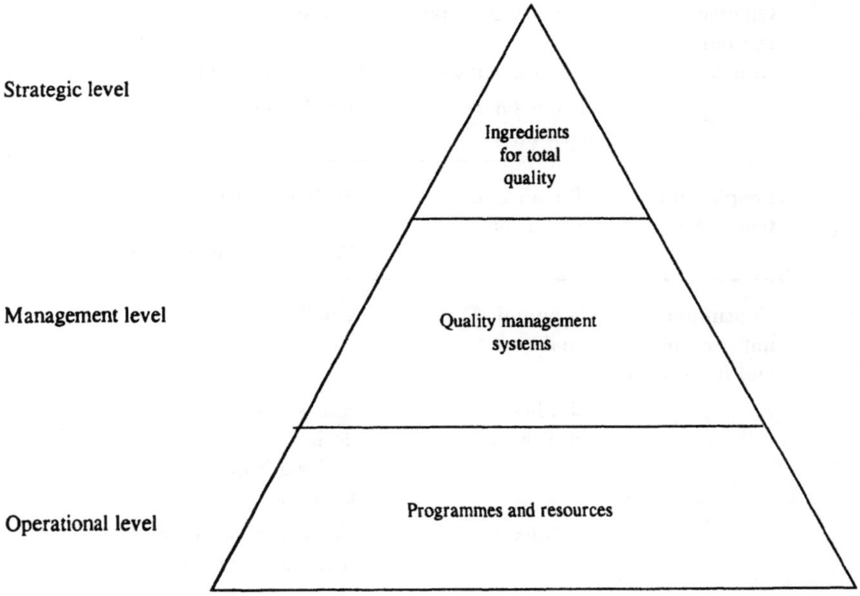

Figure 2.4: A model for total quality.

In the case of SPC implementation these three fundamental levels are:

The Strategic Level At this level, the direction (the policy) is plotted whereby an operations management based on SPC can be attained. This means that the business adopts the following concepts: all work is carried out in a system of mutually connected processes, all processes show variation, and reduction of variation leads to quality improvement. Four important ingredients are necessary: leadership; quality and customer satisfaction; people and teamwork; improvement and innovation (see also the first column of Frame 2.1). These ingredients were also seen in Deming's 14 points. SPC cannot take off without a programme which takes these ingredients into account.

Frame 2.1 Survey of the Ingredients of a Model for SPC

Strategic level	*Management level*	*Operational level*
Leadership	Realization and understanding	Various media and interactive discussions
	Change of culture	Direction and outlook
	Organisation for improvement	Quality improvement studies (PDCA cycle)
	Reward and appreciation	Money and pat on the back
Quality and customer satisfaction	Quality at supplier	Partnership
	Process control	Statistical approach
	Focus on the customer	Partnership
People and team work	Training and education	Statistical thinking
		Personal development
Continuous improvement and innovation	Economics of quality	Quality costs
	Quality improvement	Statistical thinking, design of experiments, and work process analysis
	Quality by design	Taguchi methods and statistical robust design
	New technology	Computers, robotics, and expert systems

The Management Level At this level, the quality management systems which ensure realization of the four ingredients of SPC are defined. The management system is the link between the strategic level and the operational level. Without this link it is not possible to transform the defined strategy into an SPC operations management. The quality management system is based on 12 elements around the strategic level (see Frame 2.1). The 12 elements in the second column of Frame 2.1 can serve as a starting point to help the organization set up a quality management system. However, each organization will need to make such a system specifically suited to itself.

The Operational Level On this level the specific activities needed to realize SPC are recorded and carried out. The necessary programmes are dictated by the quality management system. Resources must be made available to realize it. Statistics will be most evident on the operational level. In the last column of Frame 2.1 a few statistical methods which can be used are given. We shall not go into details in this section. For a number of techniques we refer to Chapters 6 to 9 of this book and to Montgomery (1991 and 1996).

The model with its three levels is thousands of years old and can be recognized within many organizations, *e.g.*, churches, banks, hospitals, businesses and institutes. Each unit within an organization often has its own ideas about the meaning of SPC. It is thus necessary to help each unit to develop a common outlook. In many cases the management will focus on the strategy, whilst the operators are almost solely concerned with the operational level. It is critical that everyone is aware of the whole model and knows how his contribution fits into the total.

It is often erroneously assumed that statistical thinking and techniques are only applicable at operational level. It has, however, been shown that if the statistical resources used on the shop floor are not consistent with the strategy of the management system, use of these resources will quickly decline and, as a result, create frustration at all levels. By definition the strategic level is more conceptual and use of statistical techniques does not go without saying. The case is different at management level. This level is the intermediary between strategy and operation. The management often uses aggregated data such as performance indicators, output figures, sales figures, *etc.*. If the management knows how to deal with variation, these figures can furnish an important contribution to quality improvement.

The focus on business processes is the central theme of SPC. The starting point is that all work can be approached as a process. It is important to realize that both repetition and variation occur in processes. In recent literature more and more concepts are encountered in which the term 'business process' is used. An example is 'Business Process Redesign (BPR), also known as 'Re-engineering of Business Processes' (Hammer and Champy, 1993). According to Blonk (1995) three levels can be distinguished in BPR:

At strategic level:

- □ 'Business Transformation';

At management level:

- □ 'Business Process Redesign';

At operational level:

- □ 'Work Process Redesign';

'Business Transformation' is a top down approach which begins with a strategic reorientation of the business. In the literature it is stated that seven of every ten attempts at this ambitious level fail to stay the course.

'Business Process Redesign' entails redesigning complete, often complex, compound business processes, such as the purchasing process, the production process and logistics. The goal is to achieve substantial improvements to the processes. This approach is also revolutionary and far reaching, requiring much—and continuous—attention from the management to keep everybody on the same wavelength and to ensure that employees do not become discouraged. The risk of stopping is high since it is necessary to go the whole way before the desired improvement of results becomes visible.

'Work Process Redesign' is aimed at redesigning existing work processes within the present organizational structure. It does not go further than the departments themselves, and the relationships with other business processes are temporarily taken as facts. On the one hand, this insular approach is controllable and manageable per department, on the other hand it leads to sub-optimization and conflict with other business processes. Positive points of this approach are: greater motivation of personnel and better results compared to those of the present processes.

In a synergetic approach in which top down and bottom up complement each other, the advantages of both approaches can reinforce each other while the disadvantages can be eliminated as far as possible. Blonk gives a reference model for this method, which leads to the desired result by means of alternation of the top down and bottom up approach, see Figure 2.5.

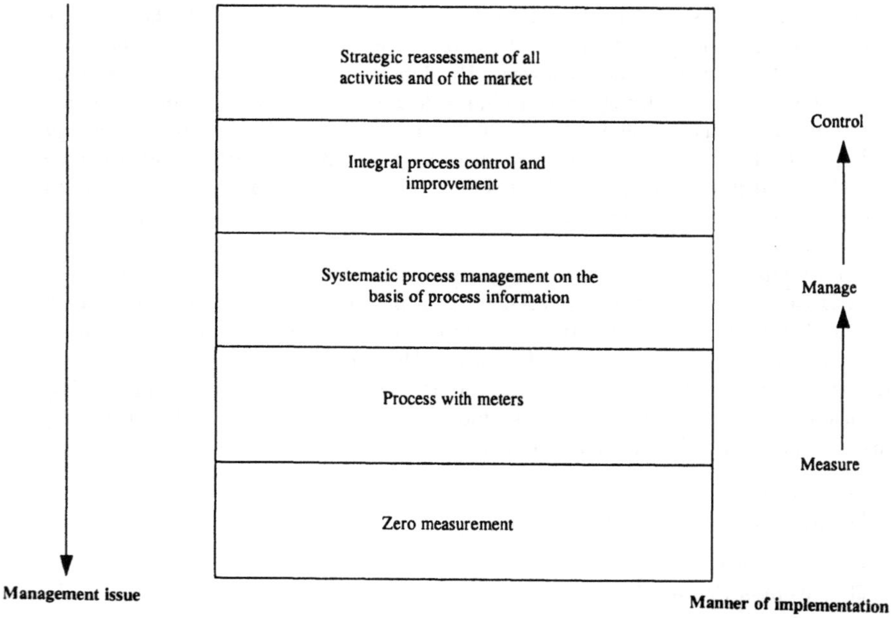

Figure 2.5: Implementation of 'Business Process Re-engineering'.

If this model is compared with the approach which we advocate for SPC, many similarities can be observed. Blonk indicates as much by stating that the application of BPR which he describes is more or less equivalent to TQM (which we call SPC and which Deming describes in his 14 points). The developmental approach advocated by SPC constitutes management of the results of business processes, aimed at complying with the expectations of internal and external customers.

2·3 Quality Assurance and Prizes

When setting up a quality system many businesses choose the series of quality assurance norms ISO 9000 (Bureau of Business Practice, 1992) as a guideline. The reason for this is that these norms are commonly accepted as a guarantee for the customer that what is being offered is of sufficient quality. The scope of the ISO 9001 norm is the most comprehensive. The ISO 9002 and ISO 9003 norms derive from it with, in each case, a more far reaching limitation of the scope. Further, without having the status of a norm, ISO 9004 gives guidelines for a quality system. Certification of the quality system on the grounds of one of the three norms follows if the business has had its system successfully assessed by a recognized certification agency.

The pursuit of ISO certificates for quality systems is unstoppable. Dutch business has thrown itself into the race. Within a few years every self-respecting business will have an ISO certificate and the possibility of distinguishing oneself from the competition will have disappeared. Of course, what remains is that any organisation which has been awarded an ISO certificate works with a good quality system.

After ISO certification the next step in quality policy of a business is the implementation of SPC. The title of this chapter: SPC as part of quality policy, indicates that SPC must not be seen as being isolated: it must be embedded in the organization. Our experience is that it is excellent if a business first sets up its organization according to ISO 9000. Work procedures, customer service, *etc.*, are then taken care of. However, implementation of SPC without an ISO certificate is also a wise management decision. SPC entails giving shape to both control and improvement of quality. But as already mentioned, an ISO certificate is a good basis for the implementation of SPC.

Progress to total quality can be guarded using the evaluation model of the Dutch Quality Prize. This model is described in full by Hardjono and Hes (1994). The Dutch Quality Prize was awarded for the first time in 1993. The prize was established by the Steering Committee for Dutch Quality (now the Dutch Quality Institute). The criteria for the prize can be used as a guideline on the road to total quality. The prize distinguishes three steps:

The first step is self-evaluation. Nine assessment criteria are used here (the contribution of each criterion to the total score is given in brackets)

1 Leadership (10%);

2 Policy and strategy (8%);

3 People management (9%);

4 Resource management (9%);

5 Process management (14%);

6 Customer satisfaction (20%);

7 Personnel satisfaction (9%);

8 Impact on society (6%);

9 Company results (15%).

These elements are ordered in the house of quality in Figure 2.6.

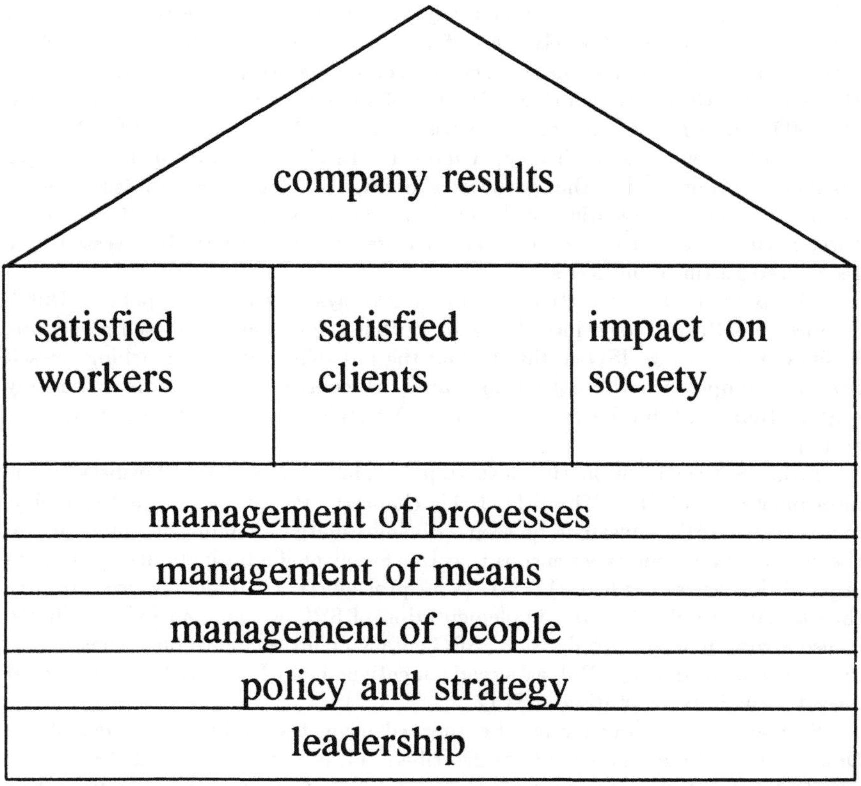

Figure 2.6: The house of quality.

Businesses which have a certificate based on ISO 9000 or which are attempting to gain one, are quick enough with answers on the self-evaluation. Note that a business which conforms completely to the ISO 9000 description will only have 375 of the maximum of 1000 points. However, if a business begins seriously with SPC, whether or not it has been awarded an ISO certificate, we dare to assert that when the SPC approach has seeped into all nooks and crannies of the company, it is possible to make a serious bid for the Quality Prize. We have experience with a company where this was indeed the case. (Philips Semiconductors at Stadskanaal was finalist in the 1994 Dutch Quality Prize.)

The second step is determination of position. The road to total quality has 5 development stages, namely:

Phase I Product oriented

Phase II Process oriented

Phase III System oriented

Phase IV Chain oriented

Phase V Total Quality

Using the answers to the self-evaluation it is possible to determine for each criterion which phase the business is in. It is, of course, important that the business assesses its own situation as objectively as possible.

The third step is to put the business forward for the prize. If the score for each criterion turns out to be predominantly in phases IV or IV, a business can apply for the Dutch Quality Prize. The self-evaluation is assessed by an experienced audit team, which judges what effort has been made to which results. Each business receives a detailed report with a description of strong and weak points and recommendations for improvement of certain aspects.

Since 1992 a business which stars in the area of total quality can be awarded the European Quality Award, the highest quality award for European business. This prize was developed by the European Foundation for Quality Management (EFQM) and the European Organization for Quality (EOQ). The EFQM was set up in September 1988 by the leaders of 14 large European companies. The goal of EFQM is to let quality play a larger role so that European industry can take in a stronger position on the world market.

The United States has the 'Malcolm Baldrige National Quality Award', which is administrated by the National Institute of Standards and Technology. The prize was set up in 1987 and it is named after the former Trade Secretary, Malcolm Baldrige. The prize is awarded annually to American businesses which have given an excellent performance in the area of quality and quality management. The prize is presented by the President.

We have already mentioned the Deming Prize in Section 1.3. The prize was set up by the JUSE and is awarded in four categories: persons, large concerns, small companies and company divisions. Since 1994, non-Japanese businesses may also compete in the last three categories.

Chapter 3

Implementation Plan for SPC

3·1 Reasons for Implementing SPC

At the end of 1994 the software company SPSS Benelux Ltd. carried out a survey together with MANS Consultants in order to obtain more insight into the problems and opposition met by businesses when implementing SPC. The survey was sent out to more than 250 companies. Of these, almost 100 sent back a completed questionnaire. Most companies were operating in the industrial sector and had between 100 and 250 employees. The respondents turned out to be mainly quality assurance or production managers between 25 and 45 years of age. More than half the companies had less than five years experience with SPC.

When the grounds for the various companies using SPC are examined cost saving turns out to be the most motivating. All of 54% of the respondents indicated that economy was the catalyst for actually switching over to an operations management based on SPC. The reason mentioned in the second place is that the customer required it (25%), while, at 17%, customer satisfaction came in the third place.

Our experience in the various companies where we supported the implementation of SPC concurs with this result. Of course, cost saving is an important motive. Implementation of SPC costs blood, sweat, and tears, and must be paid back by a multiple of the invested costs. Furthermore, customer requirements form a strong incentive for implementing SPC. With its semiconductors Philips is an important supplier in the automotive industry. The requirements of the car industry are very stringent (see the requirements of the Automotive Electronics Council (1994) in which the requirements of car manufacturers Chrysler, Ford, and General Motors are formulated). SPC is a compulsory requirement. Companies such as Fokker and Hollandse Signaalapparaten must meet the current requirements in the aviation industry and the weapons industry, respectively. SPC is indispensable hereby. Especially do American customers form an important motive for implementing SPC, because they see SPC as a minimum requirement. This occurred at ASM Lithography at Veldhoven when an American customer asked questions about SPC. Although the questions were answered as correctly as possible, the customer made a note in irritation: 'Does not use SPC'. For ASM Lithography this was the moment to decide to implement SPC.

Other reasons mentioned in the SPSS/MANS survey for switching to SPC are: building up an image as a quality supplier, and requirements from executive management that SPC is implemented. Issues such as liability and legislation played only a marginal role. 70% of the respondents reported that middle management and quality assurance personnel had participated in training in the methods and techniques of SPC. 36% reported that shop floor personnel had also been trained in SPC.

Of the businesses which were working with SPC, more than half said they were satisfied to very satisfied with the results which had been achieved. On the other hand, a quarter were unsatisfied to very unsatisfied, and 20% had no opinion. The most important bottle necks encountered by these companies when implementing SPC and applying SPC were:

- Not being aware of the possibilities of SPC;
- Not having enough time;
- A low degree of motivation and involvement in general;
- No support from the management;
- Continuity;
- Lack of assurance;
- Lack of cooperation between teams.

In this book we indicate how these bottle necks do not need to form a deterrent. The experience we obtained at the various companies ensure a practical approach which is able to form the start of an operations management for continuous improvement.

The companies which were not yet working with SPC gave as the most important reason for not doing so that their process was not suitable for SPC. In our opinion this is nonsense, as already stated in section 1.1. Every business can profit from SPC. The way in which SPC can be implemented is described in the following chapters.

That taking measurements did not fit into company culture was given as a second reason for not yet having implemented SPC. Lack of good advice and the right resources both came in the third place. This book fully meets with these latter needs.

Success stories about the implementation of SPC are scarce. On the basis of our own experience we deduce the following reasons why the implementation of SPC meets many problems in practice:

1 SPC is a question of many years.

2 SPC requires the continuous attention of the executive management.

3 SPC calls for delegation of tasks, responsibilities and qualifications to the lowest possible level.

4 SPC must be supported by someone with a thorough knowledge of what can and can not be done with statistics.

5 The will of the organisation to tackle problems on the basis of data must be discernible on a day to day basis.

6 A project management approach in teams must be the norm.

A number of these points has already been discussed (point 2 in section 2.2. and point 5 in section 2.1), or will be treated in detail in later chapters (point 1 in Chapter 10 and point 6 in Chapter 4). We will now concentrate on the third and fourth points.

The third point entails qualifications, tasks, and responsibilities being delegated to the most logical place. This appeals to Maslow's Needs triangle; see Figure 3.1 (In 't Veld, 1988).

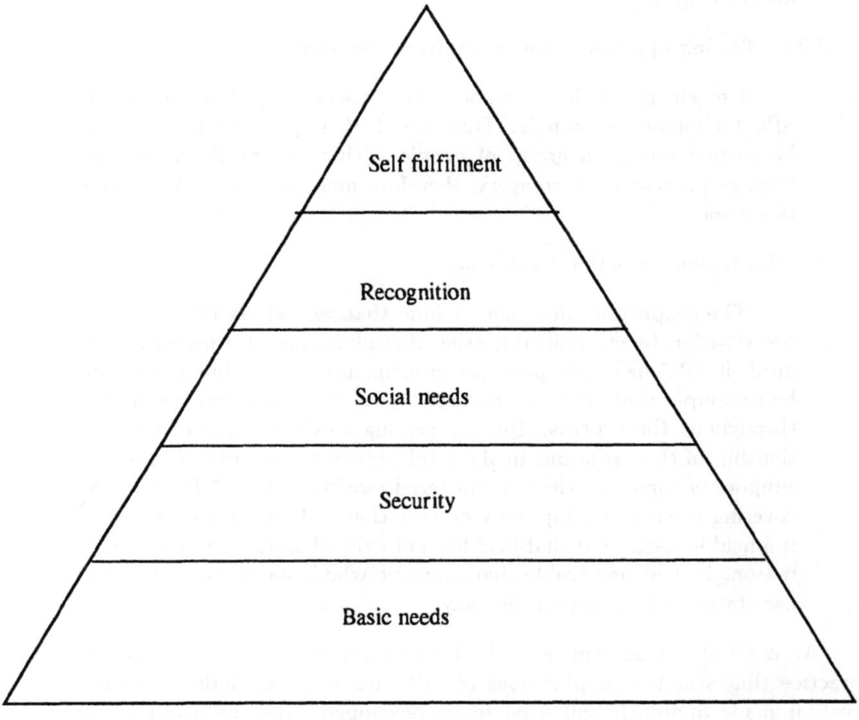

Figure 3.1: Maslow's Needs triangle.

The basis of this triangle (the bottom layer) is the primitive need of staying alive, the second layer is security, the third, social needs, the fourth, recognition, and the peak of the triangle is self-fulfilment. Transition from a lower to a higher layer is only possible if all requirements of the lower layer have been fulfilled.

By means of SPC we try to supply the individual with opportunities for self-fulfilment so that he will achieve the highest layer. The management must create the necessary conditions. If the tasks, responsibilities, and qualifications have not been properly regulated, use of the control chart is only outward show. With SPC, operators must let processes run according to their possibilities. They must use control charts and out of control action plans to trace special causes and to bring the process under control again. An operator is thus not only an extension of the control chart who, at a signal, calls someone in to remedy the problem, he is expected to take initiative to improve his part of the process.

The fourth point states it is essential that someone with a thorough knowledge of statistics contributes to supervision of the implementation. During a forum discussion at a symposium at the University of Amsterdam on the theme 'Statistical Quality Control in Industry', the following proposition was defended by Dr. J.P.M. de Kroon, a director at Philips:

> "Implementation of statistical process control requires the supervision of a professional statistician who, using fundamental knowledge, skill, and experience, is able to develop custom made solutions and to help implement them."

The following opposition was given from the floor:

> "Knowledge of the manufacturing process is just as important. SPC techniques are simple. Therefore, SPC implementation can also be carried out by a group of people with a technical background. Change processes are complex, therefore more is needed than statistics alone."

Dr. De Kroon's reaction to this was:

> "The proposition does not exclude that, as well as a fundamental contribution by statisticians, other disciplines are also necessary. Indeed, if SPC has been properly implemented the techniques should be so simple that they can be used by all operators and technicians throughout the process. But ... getting to the bottom of the relationship of the variations in the total process necessarily requires the support of someone who has mastered statistics. Why? Because discovering the relationship between variation and its generating factors is a field in itself, and that field has not existed all these years without reason. It is inconceivable that someone who is not a statistician can use statistics in a responsible way.'

We are fully in agreement with this point of view. We have discovered in practice that standard applications of SPC are very rare indeed. In general, a custom made approach will need to be developed, whereby statistics plays an important role. It is precisely the contribution of a statistician which can make the difference between successful and unsuccessful implementation of SPC.

Apart from that, we do have some comments on the contribution of the statistician (cf. also, Snee, 1991). In our opinion the statistician should preferably work as a supervisor in multi-disciplinary teams. In the team he will generally have a structuring role and he will assist with drawing up a statistical model. Using this model as a starting point, and with his knowledge of an extended scala of statistical methods and techniques, it is his responsibility to provide the statistical basis. We will even go so far as to state that the more the statistician succeeds in attaining the role of supervisor and initiator instead of that of calculator, the better he is able to carry out his job.

To conclude this section we will give a short summary of the advantages of implementing SPC:

1 **Reduction of waste:**
From a financial point of view, implementation of SPC is profitable. Even if the processes have been used for years, PATs (Process Action teams) will achieve an improvement of profits which can run to hundreds of thousands Euros per year.

2 **Predicable process**
Predictability is absolutely necessary in order to fine tune throughput times and for determining throughput times. This is obtained with SPC. 'Just In Time' (JIT) receives the opportunity of working properly.

3 **Better harmony in customer/supplier relations**
By this, we refer on the one hand to the internal customer/supplier relationship: SPC forces a business to cooperate both within and beyond department boundaries and to pay attention to what is required by other parties. On the other hand, a new sort of relationship will arise between the business and its suppliers and external customers. With SPC, the company will switch to partnerships.

4 **Insight into process performance**
Differentiation between variation owed to special causes and variation owed to common causes will first need to be made. Examination of the specifications will then point to the weak spots. The (usually scarce) resources can then be allocated to the biggest bottle necks.

5 **Clarity in process workmanship**
Apart from financial gain, immaterial gain is not a negligible factor. The core of the SPC approach is that operators are given more qualifications, opportunities and knowledge with which to carry out their work. They are allowed and enabled, even obliged, to intervene in their processes when quality is at stake. They also have the knowledge and skills to solve problems themselves according to a scenario developed by the PAT. Furthermore, operators have the opportunity to make structural improvements to the process' step.

Many well functioning SPC systems have failed as a result of diminishing attention. It is, therefore, of the utmost importance that SPC becomes deeply embedded in the organization, that it becomes a part of everyone's method of working. SPC must have a structure built up which promotes continuity. The consequence is that, at all costs, the management must continue to persevere on the chosen course.

There are, happily, internal influences which help. Naturally it is stimulating that customers continually desire ever better quality. Focusing on the customer is a good development. Finally, we advise that the company has its own source of statistical knowledge (or has easy access to external capacity), a person who also forms the 'SPC conscience', and who is also aware of the most recent developments in the fields of business and industrial statistics.

3·2 The Implementation of SPC in Four Phases

In section 2.2 we briefly looked at the role of the management. It is clear that the executive management creates the necessary conditions for SPC implementation. Without this support implementation is doomed to failure. The beginning of SPC implementation is thus characterized by orientation towards this subject. The executive management must learn the basics of SPC, and must be aware that SPC can give an important contribution to the profitability of the company. No self-respecting company will start with the implementation of SPC without a cost–benefit analysis. External influences, such as the insistence of a customer that SPC be implemented, may play a role in the discussion.

In general, executive management has only a limited amount of time available for this orientation on SPC. For this to be successful, preparations must have been made, preferably by the general manager, although part of the preparations can be delegated (*e.g.*, to the quality assurance manager). External consultants with a large amount of practical experience can also be of help here. If the executive management is enthusiastic about SPC we recommend the following project plan:

3·2.1 Phase 1: Awareness

On the basis of preparatory work which has already taken place within the company, an awareness day for the staff is a good start. The aim of this meeting is to let the staff become acquainted with the background of the SPC approach and the impact which such an approach has on the total operations management. The following subjects should be discussed at this meeting:

1 SPC entails a shift from inspection to prevention.

2 SPC is a new type of operations management whereby tasks, responsibilities and qualifications are assigned at the lowest possible level within the organization.

3 SPC is a way of establishing the capabilities inherent to processes.

4 All processes show variation, and dealing with variation is one of the central points of SPC.

5 Recognition, quantification, control and reduction of variation is thereby necessary.

6 A project management approach in teams is the norm.

The awareness day should be organized by someone with experience in SPC. It is important to realize that this day constitutes the formal start of SPC implementation. The first impression of SPC given to the management is extremely important. This impression will pursue the company during the whole initial period. Making mistakes and/or blunders is never sensible, but at this stage it can be fatal!

An alternative introduction to SPC would be that the staff visit a company which has been using SPC for some time.

Further, in this phase a few staff members and/or employees will draw up a plan for the integral implementation of SPC. Depending on the type of company, a number of discussions with staff members (possibly also in different locations) will need to be carried out in order to obtain insight into the requirements and the problems involved in an operations management based on SPC. Often some days are needed to form a clear picture of the most suitable way to implement SPC. This survey must lead to specific proposals.

3·2.2 Phase 2: Pilot Projects

After the awareness day the green light is given for actual implementation (phases 2 and 3). The most logical place to start is Production, since this is where the products are made by which the company earns its money. Furthermore, in general, production processes are split up into process steps in quite a structured way. Usually, however, tasks, responsibilities, and qualifications do not yet conform to the SPC method. Other good reasons for starting SPC implementation in Production are: discussions with R&D on not yet fully developed processes, too much variation in supplied raw materials, a lack of work procedures, and the necessity of constantly having to put thing right. Implementation should be carried out by means of sub-projects, for which an SPC Steering Committee has the final responsibility. The Production Manager is the chairman of the Steering Committee. The managers of the R&D (Research & Development), Purchasing, Quality Assurance, and Maintenance departments should also participate in the Steering Committee, which should further be completed with an SPC expert. We will take a closer look at the tasks of the Steering Committee in the next section.

In this phase of the SPC's implementation, the SPC Steering Committee will form a few pilot teams (selected on the basis of the preliminary survey), each of which are given the task to control a process step. Pilot teams are necessary to let the organization become acquainted with SPC methods. For the implementation of SPC on the shop floor, a pilot team consists of about 2 to 5 operators (at least one from each shift), the supervisor, a process engineer, a mechanic, an R&D employee, a quality engineer, and an SPC expert. The SPC expert is someone who knows all about SPC techniques and about the workings of an SPC team (which we call a Process Action Team, PAT) and who keeps track of the agreements and results of the team. Prior to the activities of a PAT, the approach is outlined in a start up course of about 3 hours. This course is tuned to the level of the operators. Further training takes place on the job. On the basis of the activity plan given in Chapter 5 the team works systematically on controlling the process step. The PAT's assignment is to:

- Describe the process step in smaller partial steps;
- Indicate which of these smaller steps are most important;
- Trace potential sources of error and variation, with the accompanying effects;
- List cause and effect combinations in order of risk of occurrence;
- Define activities which reduce the highest risks;
- If possible, define measurements which can be used as control parameters, and implement a control chart;
- Take measures to assure control.

When a PAT has gone through this whole programme and the measures have been implemented, we can speak of an operational SPC point. The time needed for a PAT to achieve this ranges from three months to more than a year. We assume here that the team meets once a week for two hours, plus an hour extra per week during the first two months when an in depth diagnosis is being made of the process in order to trace potential failure modes.

Organization of an SPC course (at Higher Vocational Level) oriented towards members of Steering Committees, process engineers, quality engineers, and development engineers can also be considered in this phase, in order to increase the acceptance of SPC within the organization. This book forms a good basis for such an SPC course.

3·2.3 Phase 3: Integral Implementation

If the first phase can be characterized as preparatory and the second phase as careful try-outs, now the time has come for integral implementation of SPC. The three companies introduced in Chapter 1: Philips Semiconductors at Stadskanaal, ASM Lithography at Veldhoven and Hollandse Signaalapparaten at Hengelo, all went through the first two phases as described above, and then decided on integral implementation. Not that all pilot teams had directly realized great savings, but because the management had acquired faith in SPC. During an evaluation meeting the pilot teams reported to the SPC Steering Committee on their activities and the results they had achieved. The impact of such a feedback is often sufficient to stimulate continuation. That savings have also been realized is a nice extra.

Integral implementation usually takes place under the leadership of the SPC Steering Committee (perhaps per company section in the case of larger companies). In Figure 3.2 the organizational framework of SPC implementation is given.

From this diagram it can be seen that the PATs are central, with a supporting role given by the Steering Committee and the executive management.

The teams as described in phase 2, the PATs, ensure control of the process' steps. When a process step has been brought under control a follow up team is assigned the task of safeguarding this control and starting a process of continuous improvement. All important process steps must be brought under control in this

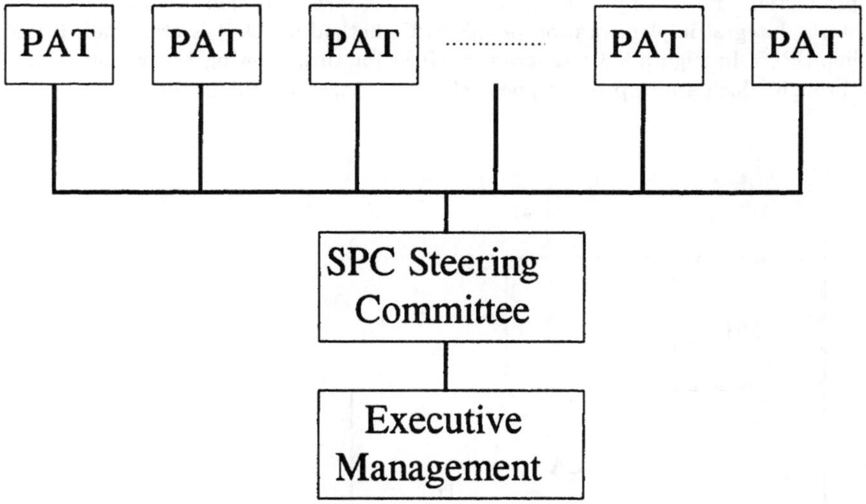

Figure 3.2: Organizational framework of SPC implementation.

way. The time needed for phase 3 is between one and a half to two and a half years.

From phase 3 onwards it is necessary that one person in the company is kept free for a substantial part of his time in order to fulfil a leading role. External assistance may be necessary both for specialised support and for supervision of the teams. In this phase the SPC coordinator and/or the external consultant should give the SPC course (see phase 2) to the whole target group (process engineers, development engineers, and quality engineers).

3·2.4 Phase 4: Total Quality by Means of SPC

It will gradually become evident that the development of new products and parts and the accompanying processes are not always optimally in keeping with SPC operations management. In the course of phase 3, implementation of Off-line SPC by R&D can be initiated. It will also become increasingly evident that suppliers must be involved in the approach. The last phase of the implementation plan is characterized by the transition to total quality. Deming (1986) shows clearly that in order to realize the implementation plan at the least all the various departments within the company must cooperate, and preferably also the suppliers and customers.

SPC implementation has important consequences for the R&D department in phase 4. By beginning with SPC implementation in Production much information is obtained about the production processes. It becomes clear which processes are in need of direct improvement and which are less critical. This leads to priorities in the order of process improvement. Besides being involved in the various PATs, development engineers are now directly involved in the SPC approach. For new designs of both processes and products the variation which can occur during production and when in use by the end user is taken into consideration from the start. Integral implementation of this SPC method in R&D is often called 'Off-line SPC'. In Figure 3.3 a diagram is given for this, showing the various stages through which a new product goes before it is taken to production.

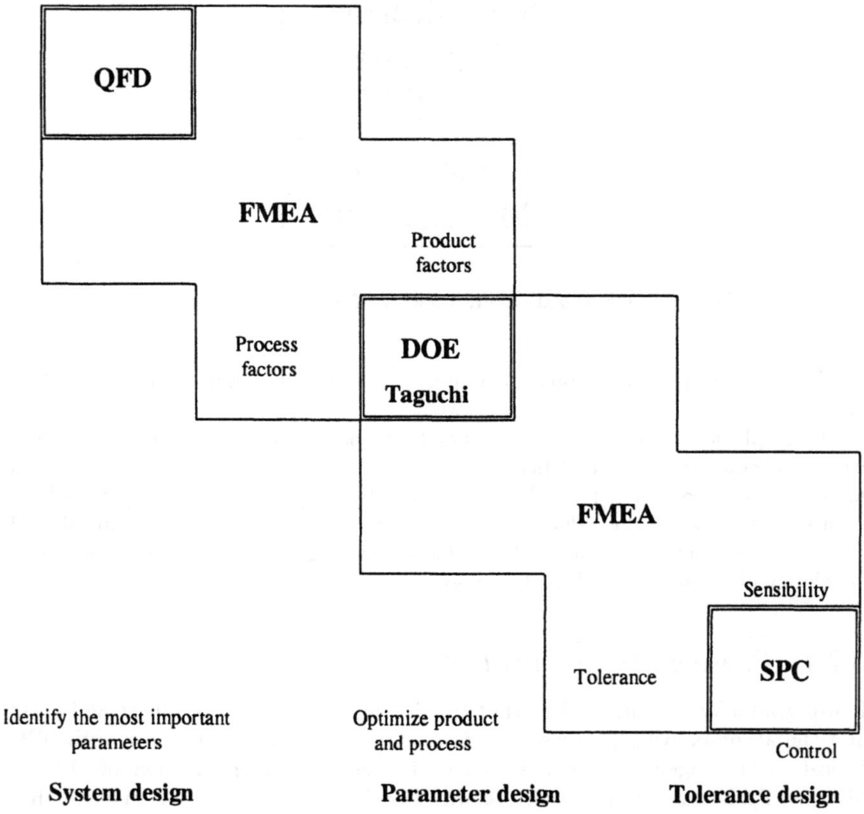

Figure 3.3: Off-line SPC.

From Figure 3.3 it will be seen that the development begins with a prototype (system design). This prototype is optimized by the parameter design. According to Taguchi's approach the design should be robust against variation during production and after. The last stage is definition of the tolerance (tolerance design), by which the various critical quality characteristics being safeguarded by means of SPC in Production is taken into account. See Hauser and Clausing (1988) for a concise explanation of the QFD (Quality Function Deployment) technique and Stamatis (1991) for the FMEA (Failure Mode and Effect Analysis) technique, and Montgomery (1991) for an extended description of DOE (Design Of Experiments).

Separate courses can be organized for all these techniques, given by teachers with practical experience.

The SPC approach in R&D is guided by a Steering Committee in which the development of processes and products is central. Leadership rests with the R&D manager; members of this Steering Committee would be the Production, Marketing, Purchasing, Quality Assurance, and Maintenance managers. Here also, the SPC expert should complete the group.

In phase 4 of SPC implementation suppliers also play a role. Now that the organization itself has gained experience with SPC over a longer period, it knows what can and should be asked of its suppliers. Experience with SPC in its own production processes has led to insight in the variation in supplied materials. Furthermore, the organization is entitled to require of its suppliers that which it practices itself. The Steering Committee responsible for SPC implementation by the suppliers is chaired by the Purchasing manager, and has the other managers as its members.

SPC can also be used for the control and improvement of maintenance. The same structure can, *mutatis mutandis*, be applied, by which the responsibility for the SPC approach in Maintenance rests with a Steering Committee.

3·2.5 Results

The time needed for the course outlined in phases 1 to 3 is about one and a half to three years, depending on the size of the organization. The following points have then been realized:

1 SPC has been accepted by the whole organization.

2 The organization is statistically oriented in its operations management, *i.e.*, decisions are taken on the basis of facts instead of on the basis of conjectures and intuition.

3 All involved have been trained in SPC and related topics.

4 As far as SPC is concerned, the organization is self-sufficient.

5 The financial revenues have become evident in lower production costs, shorter throughput times, and in a reduction of waste and rework.

6 Personnel on the shop floor have become more independent and confident.

7 The fine tuning between customers, product and process development, and suppliers with respect to producibility, specifications and dependability of supply has been improved.

At least another five years must pass before the end of phase 4 has been reached. If the planned course is followed the National Quality Prize can be applied for during phase 4. A very high placing should be achieved.

3·3 The SPC Steering Committee

The SPC Steering Committee is a group of staff members which coordinates the SPC implementation and which reports directly to the Executive Management. The tasks of the SPC Steering Committee can be summed up in four main elements:

- ☐ Steering
- ☐ Creating suitable conditions
- ☐ Promotion
- ☐ Reporting

1 Steering

1.1 *Setting of objectives and establishing the milestones.*
The SPC Steering Committee ensures that the structure necessary for SPC implementation is created within the company. The ultimate goal of SPC is to reduce the percentage of waste, to achieve predictable processes (in order to attune delivery times and throughput times), to attune internal and external customer/supplier relationships, to scrutinize the performance of the various process steps fully and finally, to ensure that the tasks, qualifications and responsibilities within the course of the process are more clear cut. This goal has been achieved when all important process steps within the company have been brought under the SPC method and a structure of continuous improvement has been introduced.

To achieve this the SPC Steering Committee sets up milestones which are periodically checked with reality.

1.2 *Determining the way in which objectives are achieved.*
The Steering Committee installs PATs, which are given the assignment of making process steps controlled. The Steering Committee provides the team with methods and resources to fulfil the assignment. The method we recommend is the activity plan given in Chapter 5.

1.3 *Choosing the process steps which qualify for SPC and indicating priorities.*
In principle, all processes qualify for treatment by the PATs. In section 1.2 we defined a process as a collection of mutually related resources and activities which transforms input into output. By 'process' we also mean, *e.g.*, the administrative processes, planning processes and financial processes. The Steering Committee chooses the steps which qualify for SPC on the basis of data concerning bottle necks in production, waste figures, implementation of new processes and/or technology (since release can only take place when the process is statistically under control) and needs/requirements of teams which are already operational (because of interconnection between various processes).

1.4 *Forming the PATs.*
The members of the team are persons who are directly involved in the daily course of the process. In the PAT the focus is on control, so that the operators (and their supervisor) are central figures. The process engineer responsible for the process step is chairman, the R&D specialist can act as secretary. The team can be further supplemented with a mechanic, a quality engineer, and an SPC expert.

1.5 *Formulating the assignments for the PATs.*
The assignments for the PATs are formulated on a special assignment form. The members accept the assignment by initialling the form. The assignment stipulates the goal to be achieved, which process is to be taken in hand, begin and end dates, reporting times, meeting times, the amount of time each member will spend, and any other special agreements.

1.6 *Determining the progress of PAT activities.*
The progress made by the PAT is reported to the Steering Committee every 6 to 8 weeks by the SPC coordinator. The way of reporting is described in Chapter 4. Possible bottle necks, necessary resources, training needs, *etc.*, are included in the report.

1.7 *Testing the results and certification of the SPC point.*
The moment for testing is decided by the PAT which, on the basis of a checklist, determines when the assignment has been fulfilled. Certification and testing is done by the Steering Committee. The department manager is chairman of the audit team, and carries out the audit together with an experienced auditor (*e.g.*, someone from the Quality Assurance Department). PAT members are present to answer questions. At the work place, verification is sought that the system works in accordance with SPC. If agreement is reached the audit team presents the PAT with a certificate which remains valid for one year. After certification, work must be made of continuous improvement of the process step. A new team called the Process Improvement Team (PIT) is installed to (structurally) improve the step (see Chapter 10).

1.8 *Setting up a control plan and keeping it up to date.*
All process steps and accompanying quality characteristics are recorded in a so called control plan. Information is given on how often, how, and by whom, measurements are taken, which control chart is used, the results of the measurement accuracy analysis, the specifications, and the result of the process performance analysis (see the following chapters for a further explanation of the various terms).

2 Facilitating Suitable Conditions

2.1 *Tracing and eliminating bottle necks.*
When implementing SPC, conflicts can arise because quantity is given higher priority than quality, because PAT meetings are not fully attended, new resources are needed, *etc.*. The Steering Committee is responsible for ensuring that these problems are solved quickly and adequately so that total stagnation is prevented. To react properly, the Steering Committee must

also make a distinction between variation owed to special causes and variation owed to common causes. If not, they are only tackling symptoms and not the root of the problems.

In section 2.2 we looked at what an operations management based on SPC means for a company. Using the principles laid down therein, the Steering Committee should ensure the main course is adhered to and make corrections where necessary. In practice this means that quality is placed alongside quantity, that priority is given to SPC, and that this is taken into account when distributing capacity and resources.

2.2 *Promoting SPC awareness.*
To obtain a good basis within the organization it is important that everyone follows the same course. The role of the management in this respect has already been described in detail. One of the tasks of the Steering Committee is to ensure that there are no doubts about the decision to implement SPC. The Steering Committee must show its engagement by, *e.g.*, showing interest in the activities of the PAT (regular visits to the shop floor really help!), propagating the philosophy (to the organization as well as to customers and suppliers) and by being both present at the start of the PAT and when certification takes place. Enthusiasm does not arise of its own accord; hard work is necessary to spark it off. An organization becomes aware that the operations management is changing when this becomes evident on various fronts. The following points can be of help here.

2.3 *Initiating training and organizing supervision.*
Training forms an essential part of the plan (see section 3.2). We have already mentioned the awareness day for the staff during phase 1 of the plan. The subject matter of this training was also discussed. For PAT members training and supervision takes place on the job while working through the activity plan. This is one of the reasons why an SPC expert should participate on the team. He can answer queries and solve problems. At the outset of the PAT, instruction lasting about three hours is advisable. During this start up it should be made clear what SPC entails, how variation can be dealt with, and what it means to make a process statistically under control. The level of the training is geared to the operators.

A more extended training will be necessary for the more specialized personnel involved in SPC implementation (*e.g.*, process engineers, development engineer, quality engineers). The content of this book would be very suitable for this training (6 to 8 meetings of about 3 hours duration should be sufficient). For further specialization, such as the robust design of processes and products, a training in the design of experiments will be necessary. Montgomery (1991 and 1996) contain the necessary theory and practical examples. For this type of training, it may be necessary to bring in an external instructor with practical experience.

2.4 *Making the necessary resources available.*
The Steering Committee reports to the executive management. The time, training and any budget for external consultants made available by the executive management for the implementation of SPC should be filled in

by the Steering Committee. In general, necessary personnel costs can be defrayed by this budget. An extra budget will often be necessary to solve bottle necks. The PATs will document their requirements (this is inherent to the methodology as expressed in the activity plan). In this way the Steering Committee can request and apportion the necessary resources on the basis of facts.

3 *Promotion*

3.1 *Promoting SPC involvement*

It is remarkable how little attention is paid to promoting results within Western European business. This is very different in South East Asia and the United States. Posters are hung on the walls, photographs, *e.g.*, of the operator of the month, SPC control charts, *etc.*. There people realize that it is of great importance to pay attention to promotion in order to stimulate understanding of and enthusiasm for SPC and to involve personnel in an early stadium. This point is further elaborated in item 3.2.

3.2 *Making promotion resources available*

Examples are:

□ Introduction of existing and new PATs on placards or in the company newspaper.

The teams can be introduced to the personnel by means of photographs and a short description. Publication should take place as soon as a new PAT has been instigated. In addition, a short explanation can be given why PATs are being set up. In this way PATs are given an identity within the company and personnel realize that the company considers PATs to be important.

□ Making PATs manifest within the departments.

PATs should be visible for personnel. Moreover, auditors and customers can see that the company is seriously working on SPC. This can be achieved by hanging a notice board with the names, aims and members of the team next to the process step. The layout of these notice boards should be similar and they should be easily distinguishable from other types of notice board. In this way PAT members can see that their work is visible, and personnel who are not PAT members can see where process control is being implemented. The boards can also display notices from the PAT, results of the activity plan, control charts with observations, the SPC certificate (if awarded) and anything else which is relevant to the PAT. It is, however, important that the information is kept up to date. An example of an SPC notice board is given in Chapter 4.

Another form of recognition is achieved by hanging SPC balls from the ceiling near the process step: red for processes which are not yet subject to SPC, orange for processes with a PAT, and green for processes which have obtained SPC certification. Incidentally, we have not yet succeeded in creating a 'ball hall' anywhere!

- ☐ Introducing milestones.
  When a PAT has achieved its goal, this should be celebrated. Not only does personnel like to see that the achieved results are appreciated, but the management also likes to express its gratitude and to stimulate the personnel to continue in this way. Besides notices on the notice board in the department, or in the company newspaper, recognition should take the form of a collective lunch or dinner, or a party, outing, group award, or gift.
- ☐ Publication of an SPC booklet.
  A booklet giving a clear description of how a company views SPC, with an elaboration on how the implementation is going to be carried out, provides the opportunity of informing everyone within the company. ASM Lithography at Veldhoven has such a booklet available in both English and Dutch (Does, Van Dooremalen and Stam, 1995a and 1995b). This is also used to inform external customers and suppliers about the SPC approach.
- ☐ Publication of an SPC bulletin.
  News about SPC concerning the PATs, the Steering Committee, the PITs, the suppliers, and external customers could be included in a bulletin. Operators can report on their own PAT. Before a company starts a bulletin thought should be given to matters such as who is on the editorial board, what the content is to be, and how continuity can be guaranteed.
- ☐ Election of the Improvement Team of the Year.
  An annual election, during which improvement teams present their approach and results to the executive management, is an effective form of promotion. In Chapter 12 we will give a more detailed description of such an election, including a list of questions for the jury. Such a competition provides the opportunity of promoting the PAT (and in a later stage the PIT). Because of the tension involved, the competitive element can be used to heighten the commitment of participants, other personnel and management. There should be good publicity about the election. The notice boards, company newspaper or SPC Bulletin can be used for this purpose. Furthermore, the election should be open to all personnel. The preparatory work and the coordination of the elections are the responsibility of the Steering Committee. The election jury consists of, *e.g.*, one of the executive managers (who is then also chairman), the Quality Assurance manager, the Human Resources manager and a representative of the employees. The prize is naturally in accordance with the importance given to SPC by the executive management. An important facet of such an election is that it is often the first time that operators present an explanation to the management of what their daily work actually entails. There have been elections at Philips Semiconductors since 1993, and this has certainly contributed to a greater familiarity with and acceptation of SPC.
- ☐ Other companies come and visit.
  When a business is well on the way with SPC, other companies become

interested. Organization of a meeting whereby other companies are invited to come and have a look at the SPC approach can have a stimulating effect. ASM Lithography, Hollandse Signaalapparaten and Philips Semiconductors Stadskanaal all have had positive experience with this sort of visit.

- □ Other means.
  Anything which increases promotion of SPC: company video, news trailer at the main entrance, interactive media, SPC slogans such as 'The PATh to improvement', 'It's worth a PAT on the back', *etc.*.

4 Reporting

4.1 *Reporting to the executive management.*
The Steering Committee is responsible for informing the executive management regularly about progress made in SPC implementation, on the aspects relevant to company policy and on the costs and benefits of the SPC based operations management. The Steering Committee draws up a quarterly status report on the progress made. This report contains the following ingredients:

- □ Progress in realization of the SPC annual plan;
- □ The number of process steps under SPC study;
- □ The number of process steps whereby SPC study has been completed;
- □ The number of PATs;
- □ The number of certified PATs;
- □ Progress of the non-certified PATs as regards the activity plan;
- □ Improvement actions as a result of PIT activities;
- □ Training a education:
  - the number of basic training sessions held (especially for PAT members);
  - number of participants of the basic training;
  - number of extensive SPC courses held (Higher Vocational Level);
  - number of participants of the SPC courses;
  - number of follow-up training sessions held (*e.g.*, course on the design of experiments);
  - number of participants of follow-up training.
- □ Costs and benefits (see point 4.3).

4.2 *Advising executive management concerning SPC policy.*
The executive management makes decisions on SPC policy and indicates how and when this policy must be put into practice. The Steering Committee reports possible disturbances and gives advice for improvement to policy. It does this by advising the executive management how bottle necks which interrupt progress can be removed. It also makes suggestions for the SPC budget and investment in resources. The more SPC implementation

progresses, the more departments which will become involved in the plan. The Steering Committee gives advice on these matters to the executive management. Finally, since the executive management stimulates SPC by means of an annual election of the best improvement team, the Steering Committee reports necessary data and coordinates the elections.

4.3 *Cost-benefit analysis.*
This analysis contains the following aspects:

- □ Costs:
  - number of hours spent;
  - investments for SPC.
- □ Benefits:
  - The PATs which wish to become certified indicate what the various activities have yielded. This also gives information on the status of the process;
  - The PITs indicate the savings and the improvements achieved by their activities;
  - Trends in reject registration gives information on cost savings.

Besides the financial gain, the immaterial gain is also important. We have already mentioned this in section 3.1 (in point 5 of the advantages of SPC implementation: clarity of process management). A company can not be managed only on the basis of tangible figures. Features such as customer satisfaction, employee satisfaction, motivation, and suchlike, are not easy to measure. Yet these aspects are also of great importance for the well being of a company.

Chapter 4

Introducing SPC with Teams

4·1 Introduction

A process is only under control if all its constituent parts are. The law of the weakest link therefore says that the process is as controlled as its weakest part is. In the previous chapter we described the impossibility of improving an out of control process overnight. For introducing SPC a four phase approach is necessary, beginning with one of the weakest process steps. Then all other critical parts must be brought under control as well, a task which may take years altogether. The introduction of SPC requires full support of the top level of an organisation, and coordinated activities of many others.

4·2 Arrangement of Many Organisations

In many organisations, both industrial and service, the separate parts of the process are performed by separate people. In industry the machines are generally operated by a small group of people (the operators), while others are filling jigs, and again other people do final inspections of the products. In service organisations some people (the so called front office) serve the clients at the counter, while others (the back office) process invoices. Modern management principles abandon such rigid organisations, which are in fact owed to the ideas of Taylor, and ask for more flexibility of the employees (see *e.g.*, Socio-technical Systems Design, Chapter 10). But even then only a few employees are closely involved at a certain process step. These people are experts, however, and know exactly how to operate that particular process step best, and they are aware of all possible problems that might be encountered as well as solutions to cope with them.

A manager or process engineer may have a good overview of the whole process, but he does not have the detailed knowledge of the operators of a specific process step. Therefore operators can really not be dispensed with when this process step must be brought under control. Operators are day to day confronted with all peculiarities of the process, they signal out of control situations and trends first, and they take corrective actions and warn the engineers or mechanics as early as possible. It is asking for trouble if operators are not sufficiently involved in the introduction of SPC—the definition of out of control situations, trends and corrective actions. A cross-functional team, with both operators and engineers, is the obvious group to bring a process step under control.

4·3 Composition of an SPC Team

The direct employees around a process step (they will be called operators in this book[1] form the backbone of a team that brings the process step under control. But they can not do that by themselves, in general, if only because they often have no power, nor opportunities of bringing about major changes. That is why active support (and not just permission) of superiors is indispensable. The best construction (see Chapter 3) is that a steering committee is responsible for introducing SPC overall, and that teams are formed for bringing separate process steps under control. Such teams are called Process Action Teams (PATs).

Which *operators* should be member of a PAT? It would be ideal if all operators who are involved in this process step participated, but this is hardly ever possible because:

- ◻ The group is too large;
- ◻ Not all operators can be spared at the same time;
- ◻ Some are not interested.

It is really important, however, that at least every shift is represented in the team. Thus it is safeguarded as much as possible that all operators are informed either directly or through a close co-worker, because it really matters that everyone knows about the agreements, plans and experiments, and that all can give their opinions and ideas. The operators in a PAT are stimulated to report the progress in their shifts. This may be unfamiliar to them, then it might be useful to give them support.

A *process engineer* or *production engineer*, the one who is responsible for the standard operating procedures for operators, needs to be part of the PAT as well. As it turns out, changes are often required. If the process step is performed on a machine, or the tools are critical, then participation of a *mechanic* is also mandatory. Moreover, it is often necessary that all kinds of maintenance activities need

[1] Titles and descriptions of functions differ from organisation to organisation, but in this book we opt for a certain system. The reader will have to translate it to his own organisation.

to be done before a process step can be effectively controlled. And, finally, maintenance is often a critical success factor to consolidate control. In all situations where communication between operators might be a problem, the ***supervisor*** (the superior of the operators) needs to be part of the PAT. His or her role is to guard the communication and to solve occasional organisational bottlenecks. Sometimes his authority is needed to force a decision among disagreeing operators. An *SPC expert* is needed in the team, certainly when the organization has only little experience with SPC. His foremost tasks are to assist in the use of methodological tools (such as the '7 tools'; Ishikawa, 1974) and to do data analyses. But no less important is that as a 'neutral' outsider he can guard the progress of the team, play a conciliatory role in conflicts, give extra 'voice' to the weakest party (often the operators), and take notes (although an external SPC expert should avoid doing this, for practical reasons).

The participation of the following disciplines is optional:

- ☐ Development—to explain the reasons behind procedures, and to do experiments;
- ☐ Purchasing—for contacts with a supplier, in case incoming material is critical;
- ☐ Quality assurance—to give a voice for the customer.

Owing to the size of the PAT we advise making arrangements about the *ad hoc* participation of these disciplines. They join the team only when their contribution is useful and/or needed, but they are not regular members.

The PAT leader must be a member of the production department. Not just his function, but also his personality is important: some people are simply better leaders than others! His most important tasks are to report to the SPC steering committee (on progress, change proposals, and bottlenecks) and to facilitate progress. If he also wants to participate actively in PAT sessions he had better let someone else (*e.g.*, the SPC expert) chair the meetings. The PAT leader takes care of regular meetings and necessary facilities.

The most important task for the PAT secretary is to take notes of agreements and the results of the PAT meetings, which in fact are the results of the activity plan. Unless the PAT has its own meeting place where walls can be used for a permanent display of the results, the notes of the secretary are input for the next meetings. Good notes are therefore essential, and should be taken by someone skilled. Developers and engineers are therefore good candidates, because they are well educated and have experience in making reports. If necessary the SPC experts should be able to take notes, but when they participate in many teams this might not be a good idea.

4·4 Assignment of a Team

A Process Action Team receives a well defined assignment of the SPC steering committee. The assignment also finalises participation of the several disciplines (production, development, purchasing, quality). The most involved member of the steering committee will be the commissioner and contact person for the team.

To standardize assignments of the teams we advise the use of standard assignment sheets (see Frame 4.1 for an example). In this sheet the following topics can be finalised:

- □ The results to be achieved;
- □ The time frame;
- □ Reporting to the SPC steering committee;
- □ The expected amount of time required from the participants.

Sometimes special arrangements are necessary, *e.g.*, about the participation of operators in shifts. An arrangement might be for the night shift to be released from participation in a meeting (to allow them to have enough sleep). Another possibility is to send such an operator home earlier, to give him enough time to sleep so that he is quite rested when at the meeting.

Often the assignment will be to analyse the process step first, then to bring it under control, and finally to secure control. The PAT has achieved its goals when there is proof that control has been achieved. The time this can take varies from one organization to another, but it depends, amongst other things, on the focus of the organization on SPC, the size of the process step, and the necessary amount of work for analysis and improvement. Often the total time is greatly underestimated. To bring a process step under control may well take half a year, but throughput times of a whole year are no exceptions. And afterwards another three to six months are necessary to implement the SPC system on the shop floor, and to prove that the process step is under control.

Fortunately, soon after the start of the implementation of SPC one may already expect benefits from improvements, even though the whole period may take about one year. And some improvements can be quite substantial. The reason is that a systematic approach is used in combination with a thorough communication structure. The improvements will show up in better operating procedures, less machine downtime, less waste of raw materials, better quality of final products, *etc.*. In the following chapters detailed examples will be shown.

The time investments of the PAT members differs. Of course everyone needs to attend the weekly meeting (2 hours per week). Time for extra activities should be added; the leader needs time for occasional reporting, and bottle neck solving (1 hour); the secretary takes notes (2 hours). We list some other common activities:

- □ operators do additional measurements, devise check sheets, and assist with experiments;
- □ the process engineer examines improvement suggestions, designs and analyses experiments, orders samples of new materials, adjusts operating procedures;
- □ the mechanic improves the equipment, orders or constructs new tools;
- □ the supervisor organizes meetings in which all operators are informed, talks with unwilling individuals, solves organizational bottle necks;
- □ the SPC expert gives on the job training on the 7 tools, and analyses data whenever the simple tools are no longer sufficient.

Frame 4.1 Process Action Team assignment sheet

1 The PAT ....
of department .... works on:
process step: ....
measurement: ....

2 The PAT achieved its goal when:

- ☐ an analysis of the process step has been made;
- ☐ an analysis of the measurement has been made;
- ☐ an analysis of the measurement variation has been made;
- ☐ a control chart with Out of Control Action Plan (OCAP) has been devised;
- ☐ the control chart and OCAP have been implemented;
- ☐ a process capability study has been made;
- ☐ the periodical evaluation of control chart and OCAP have been organized;
- ☐ ....

(tick the appropriate tasks)

3 Starting data: ....
Estimated end date: ....
Estimated certification date: ....

4 The PAT reports the progress to the SPC steering committee at least every ..... weeks.

5 The PAT has a meeting every: ....

6 The amount of time required from each member will be about ..... hours per week.

7 Special arrangements: ....

Drafted by: .... initials: ... date......
Agreed by PAT Leader: .... initials: ...
Agreed by PAT Member: .... initials: ...
.... initials: ...
.... initials: ...
.... initials: ...
.... initials: ...
.... initials: ...
.... initials: ...
.... initials: ...
.... initials: ...

The time spent on these activities may add up to some hours per week for all members of the PAT.

Many organisations make use of a *suggestion box* system. This is meant to stimulate operators to find improvements. A successful idea is rewarded. In these organisations the introduction of Process Action Teams requires arrangements in dealing with the suggestion box system. Obviously it is not the intention of creative thinking within a PAT that ideas are found in the suggestion box afterwards. A possible approach to this problem (it is a problem indeed, because creative operators may earn quite a bit of extra money with the system) is to consider the PAT minutes as proof of the suggested improvements, and to reward ideas not directly related to the assignment. Suggestions from others than PAT members could be judged by the team, and rewarded if they are useful.

4·5 Meetings of the Team

In the first meeting the members of the PAT are extended a hearty welcome by a closely involved member of the SPC steering committee, who will also stress the importance of process control and the role of the team. By his personal presence he also underlines his own involvement. After this introduction the team gets a short SPC course. The teacher (most likely the SPC expert) spends some 2–3 hours on the following topics:

- ☐ What is SPC;
- ☐ Dealing with variation;
- ☐ The control loop and control chart;
- ☐ Assignment of the team and the activity plan.

The Process Action Team can then begin its work.

From the viewpoint of 'rhythm' we advise to meet regularly, at a fixed time (*e.g.*, each Thursday, 3 to 5 pm). In the beginning more meetings are necessary than after a while; meetings every fortnight may well be sufficient once the train is moving well; an organization with a high focus on SPC might instead start with two meetings per week. For operators on shifts special arrangements may be needed to allow for their presence or to legitimize their absence. The example of an operator in a night shift has already been mentioned, but arrangements are also necessary for operators who need to interrupt their shift system's 'week-end' in order to attend a meeting. A solution might be that such hours may be compensated with a factor of 1.5.

The agenda of the PAT meetings is dictated by the *activity plan* (explained further in the next chapter).

1 Process description

2 Cause and effect analysis

3 Risk analysis

4 Improvement actions

5 Measurements

6 Measurement analysis

7 Control chart

8 Out of Control Action Plan (OCAP)

9 Process Capability Study

10 Certification

Moreover, the progress of improving actions will be reviewed.

Meetings take at least $1\frac{1}{2}$ hours, but more than $2\frac{1}{2}$ hours is often no longer efficient. Sometimes it might be useful, however, to spend at least half a day on an extensive task (*e.g.*, a complete cause and effect analysis).

4·6 Communication by the Team

In its assignment to the team the SPC steering committee also sets the frequency of progress reviews, which is a task for the team leader. In addition, the steering committee needs to know the progress of all PATs. To this end the progress sheet of frame 4.2 has been designed.

At least as important is the communication of the PAT with the other people involved in the process step. If for one reason or another (and there are many!) they withdraw from the appointments made by the PAT, then the mission of the PAT—to bring the process step under control—will certainly not be achieved. Good communication is an important first step in overcoming resistance and opposition. It is useful to have the work of the PAT on the agenda of the regular meetings of operators. A member of the PAT reports on the subjects that are worked on, the appointments and the actions. When the PAT has done a lot of work, it is a good moment to write a booklet with a summary of all activities and decisions of the team. This booklet can then be distributed amongst everyone who is involved with the process step—all with the aim to share the knowledge among all relevant people.

But communication on even shorter notice is also necessary. The operators in the PAT must be stimulated to discuss the results of the PAT meetings immediately with their colleagues. And the use of a *PAT notice board* is strongly recommended. Every PAT should use its own board to show all relevant information. A standard notice board for SPC within the whole organization helps to make the subject more recognizable. A PAT should use this standard, and tailor it to its own needs:

- ☐ A presentation of the team and the process step;
- ☐ The results of the activity plan;
- ☐ The improvement actions (planned and finished);

Frame 4.2 Progress Monitoring Sheet for PATs

Date 07/16/1993 Department	Process Step	Leader	Start Week	Finished Step 1	2	3	4	5	6	7	8	9	10	End Week
Diffusion	Water Grinding	Beuker	9237										•	9326
	Sand Blasting	Koops	9306						•					9406
	Metal Evaporation	Van der Laan	9318			•								9413
Glass Bead	Glass Deposition	Hemsens	9306			•								9348
	Silver Evaporation	Strik	9326	•										9420
	Measuring / Selecting	Zwiers	9340											9427
Implotec	Tape Sealing	Meedendorp	9306										•	9326
	Crystal Etching	Van Dijk	9306							•				9345
	Tinning	Venema	9338											9440
	Imploding	Hogenberg	t.b.f.											
Small Signal	Tinning	Groenewold	9237									•		9404
	Imploding	Oost	9304							•				9339
	Measuring / Selecting	Van der Kaap	9330		•									9351
TOTAL				1	1	2	-	-	1	2	-	1	2	

☐ The status of the actions, and remarks.

Because all boards have the same layout (see Frame 4.3 for an example), they are easy to recognize for both employees and visitors. Together they contribute to an impression of SPC that can only be positive, also from a PR viewpoint (see section 3.3). For a well functioning notice board the PAT members have to appoint individuals to maintenance it: who is responsible for what? If this is badly organized the notice board will work against the team! The damage to SPC through a neglected PAT notice board (where the most recent information dates from a few months ago, *e.g.*) can hardly be overestimated.

4·7 Winding Up the Team

The team will be released from its assignment when the goals are met. On the occasion of this the SPC steering committee should congratulate the team and reward the members for their efforts and results. A good sized piece of pie is just the bare minimum.

Frame 4.3 PAT Notice Board

The PAT notice board has three goals:

- ☐ Communication;
- ☐ Recognizability;
- ☐ Public relations.

A standard with a fixed format is therefore very important. Figure 4.1 is an example of such a board. It is divided into four parts:

upper left side: presentation of the team and the process step;

lower left side: the assignment and background information;

upper right side: the status;

lower right side: improvement actions and results.

Of course, any other well considered layout is as good as the one above.

To check whether the mission of the team is completed, a checklist is a useful tool (see *e.g.*, Frame 4.4). Organizations using a standard assignment sheet will have more or less equal assignments, so checks are made really easy using such a check sheet. It may be advantageous if the quality manager of the organization assists with the evaluation. He or she represents the customer, is responsible for uniform standards (in larger organizations), and has experience with audits.

With the help of the checklist the following topics are evaluated:

- ☐ Are the elements of SPC present (control charts, OCAPs)?
- ☐ How do the operators use the SPC elements, and what is the supervision like?
- ☐ How effective is the process?
- ☐ Is the documentation up to date (FMEAs, operating procedures)?
- ☐ How will SPC be maintained (and improved) in the future?

A successful team can be rewarded with a certificate, an official document proving that the mission has been completed. Two examples are shown in Figure 4.2.

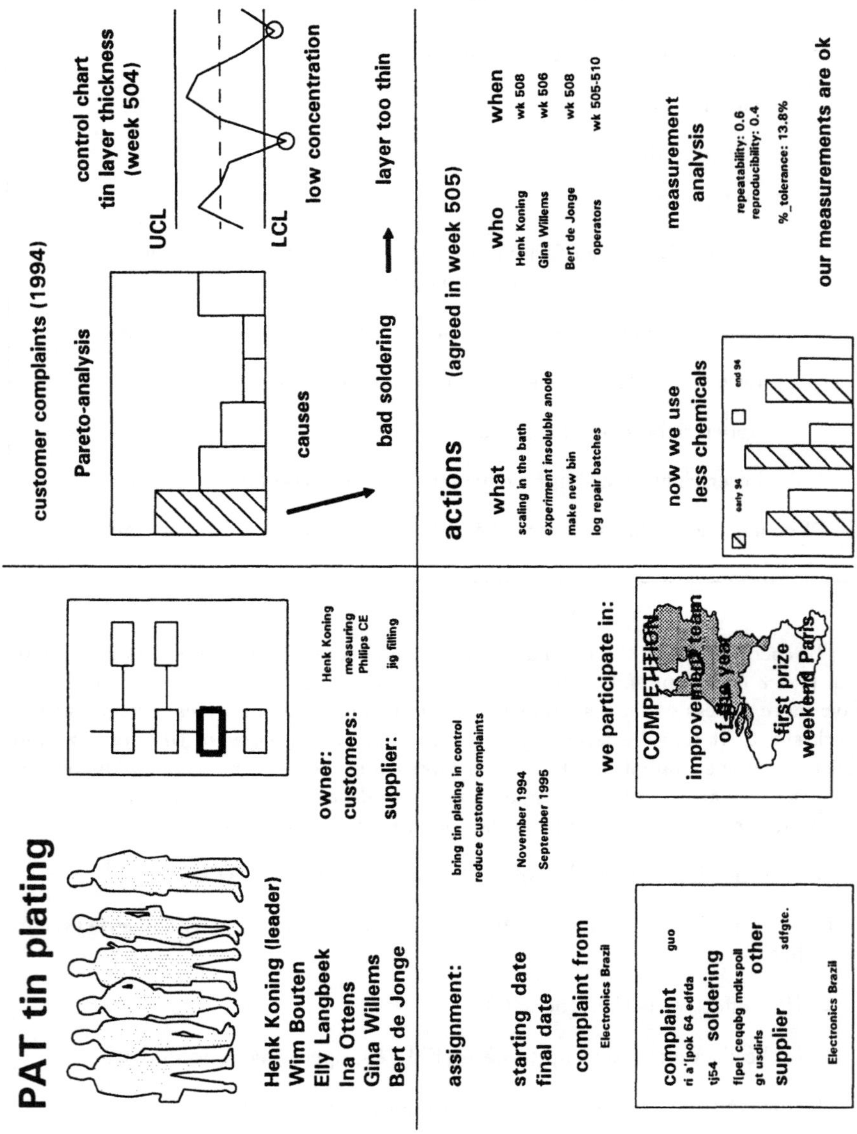

Figure 4.1: Example of a PAT notice board.

Figure 4.2a: Example of certificate.

Frame 4.4 Checklist for the Evaluation of a Process Step

1 *Presence of SPC elements*

- 1.1 Is the process step monitored through control charts?
- 1.2 Are the control limits correctly evaluated?
- 1.3 Are there Out-of-Control Action Plans?
- 1.4 Are out of control situations logged?
- 1.5 Are the specification limits well defined?

2 *Use of SPC elements*

- 2.1 Do all operators fill in the control charts correctly?
- 2.2 Do all operators recognize out of control situations?
- 2.3 Do all operators know the difference between a control limit and a specification limit?
- 2.4 Do all operators use the OCAP in an out of control situation?
- 2.5 Are out of control situations correctly logged by all operators?
- 2.6 Is process information passed on at shift changes?
- 2.7 Does the supervisor check whether control charts are filled in?

3 *Capability of the process step*

- 3.1 Is the process step under control?
- 3.2 Does the process step meet the specifications?
- 3.3 Is the measurement capability sufficient?

4 *Documents control*

- 4.1 Are operating instructions conform to the arrangements made about control charts and OCAPs?
- 4.2 Is the FMEA of the organization updated?

5 *Maintenance of the SPC system*

- 5.1 Is there a team responsible for maintenance of the SPC system?
- 5.2 Are control limits reviewed regularly?
- 5.3 Are the OCAPs evaluated and improved regularly?
- 5.4 Is the process capability regularly evaluated?

4·8 Assurance of an SPC System

It is a serious mistake to think that an under control process will remain so forever. To secure such a result measures should be taken by the team so that future attention is guaranteed. That is why the checklist of Frame 4.4 has a paragraph devoted to the maintenance of the SPC system. In Chapter 10 the subject of maintenance will be dealt with extensively.

Figure 4.2b: Example of certificate.

Chapter 5

The Plan of Action for Introducing SPC

5·1 Introduction

In this chapter we describe the ten steps of an activity plan that ultimately yield a controlled process. Each Process Action Team (PAT) passes through these steps, so that it is always clear what subject the team should work on. Moreover, the commissioner and SPC steering committee have a simple tool to monitor the progress of PATs (see Frame 4.2). Another big advantage is that a common language develops—everyone knows what is meant by *e.g.*, a 'cause and effect analysis'. In the scheme of Figure 5.1 an overview of the 10 steps is given, together with an indication of the throughput time.

Most of the illustrative examples stem from the semiconductors industry, more specific from manufacturing diodes. The advantage for the reader is that a diode is a relatively simple product, and the production process consists of only a few vital parts. It gives the authors the opportunity to show the wealth of SPC problems within one organization. Diodes are high volume products. At the end of this chapter we will also pay some attention to low volume processes, because then the activity plan might need adaptation.

Diodes (electrodes with only two connections: 'di-odes') are small components, conducting an electrical current in one direction and blocking it in the other. They are use on printed circuit boards as switches or rectifiers. It is a typical high volume product, manufactured partly in physical/chemical processes and also party by assembly. Efficiency is of utmost importance because the competition from low income countries is fierce. In Figure 5.2 an example of a leaded glass diode is shown.

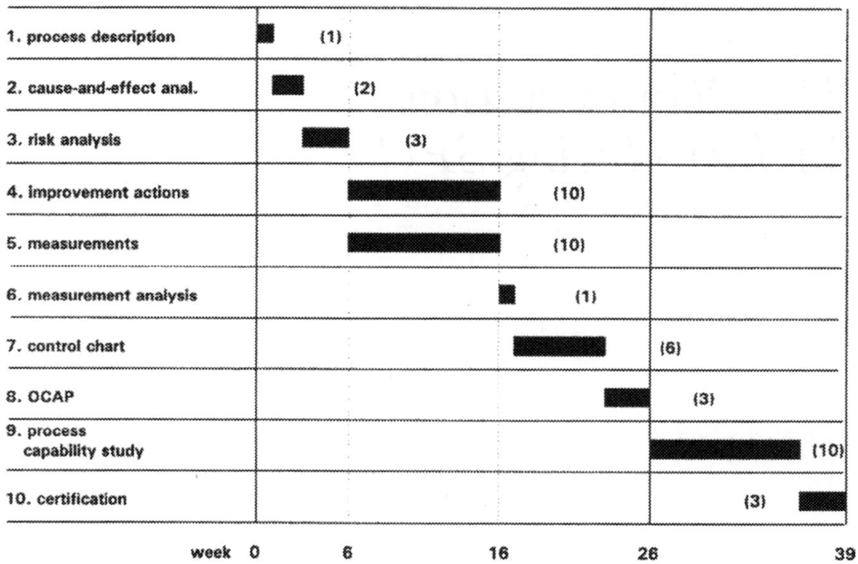

Figure 5.1: The activity plan.

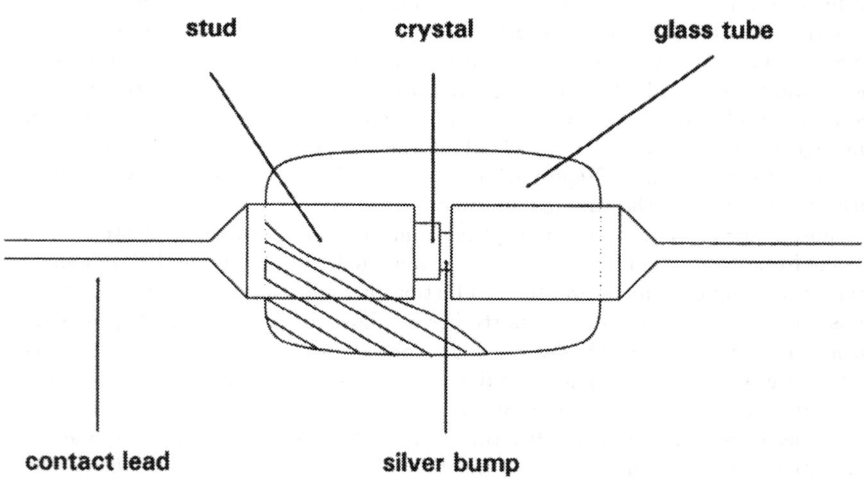

Figure 5.2: A leaded glass diode.

5·2 Process Description (Step 1)

1 Summary

The process description is the first step in the analysis of a process step. Beginning and end are clearly defined, and everything in between is described in small parts. The steps with an important contribution to the final result of the process step are indicated.

2 Goal

The PAT members have to define exactly what part of the process is their responsibility. Everything that happens before is the responsibility of the 'supplier', and the 'customer' (the following process steps) is responsible afterwards. A clear distinction of responsibilities is important because the team should be able to concentrate on its own process step, and not be bothered by problems it cannot really solve. Division of the process step into small parts makes all activities to be done clear, and in which order. This exercise may show that different shifts have different ways of working; or it shows that theory (work instructions) and practice differ. To agree upon the vital activities of the process step the most important parts of work have to be indicated.

3 Result

The process step is well described, conforming to reality. Different ways of working have been discussed, and maybe a single 'best' way is selected. Everybody agrees on the heart of the process step. This result is written down on the process description sheet (see below).

4 Method

All parts of the process step, from beginning to end, are summarized. This is done by mentioning the activities, so the description starts with a verb (*e.g.*, 'set up machine' or 'check balance'). A process description is a (simple) flow chart in words, without the corresponding symbols. The parts are numbered for later reference, and the starting position is given the number 0. The normal situation is the reference; special cases are considered only if they bring about special situations. If, *e.g.*, a machine is started up on monday morning and remains operating the rest of the week, then the situation with a running machine is normal and should be described. Unless, of course, it is known beforehand that this Monday morning effect is an important cause of trouble—it would be foolish to neglect this fact. The final step in the process description is to indicate for every activity whether or not it is important for the result of the process step.

5 Guidelines for Use

Often some kind of process description exists already before the team starts. For the aim of process control this description is in general not appropriate, however, because it is too technical, originating from a process engineer or developer. Moreover, it states how the work should be done instead of what really happens—and this friction is often a cause of trouble. Therefore the team describes the process step in its own words. Based on this summary, discussion about the reasons for deviating from the operating instructions can be fruitful. Some common reasons are:

1 Operating instructions may not have been updated for several years and no longer fit with reality. Sometimes there are no instructions at all, and operators have developed their own operating procedures.

2 It may happen that there is so much work to do that operators simply cannot do every prescribed check. This will become the standard when the supervisor allows for this neglect. In this situation the possible effects must be reconsidered. If the check is nonetheless very important, then opportunity should be created, otherwise it is better to cancel it officially.

3 Operators often find more clever ways to do the job. The PAT then considers advantages and disadvantages and comes to agreement about what is best. Should this not be possible, *e.g.*, because two opinions cannot be reconciled, then further data collection and analysis might be necessary. But sometimes there is not a single best way; then it is as well to establish this formally. This may be the case when the design of a work station forces tall people to work differently from their smaller colleagues.

4 Practices may have grown without anyone being able anymore to explain the reasons behind them. When the importance of such activities are judged, these practices will surely be discussed. The knowledge of a developer or quality engineer is needed to decide upon the correct operating procedure.

The amount of detail of the description is a difficult matter. It might be useful to see the description in terms of 'levels'. On the first level the process step is divided into about five smaller parts; one level deeper splits every part up in about five smaller parts, etc. Mostly a description of two levels deep is sufficient in step 1. Later on in the activity plan a particular part might need more detailed description, but only if the part is very important and the root cause is sought.

6 *Case Illustration*

The electrical characteristics of a diode are mainly determined by its semiconducting crystal. These crystals are produced with a diffusion process in which silicon wafers are 'contaminated' with phosphor and boron atoms. From one wafer thousands of crystals are assembled. Many crystals from the outer edge of a wafer may be of inferior quality. The crystals are then assembled into diodes, but during the assembly process the quality of the crystals might be badly affected. For these reasons all assembled diodes are subjected to electrical tests. High quality products are selected as 'main type', some lower quality products are selected as 'sub type', and all inferior products are scrapped.

A Process Action Team got the assignment to examine the measurement/selection process. The team chose the epithet "Measure it well, know it better" (translated from "Goed meten, beter weten") to express its aim: make certain that the product is measured correctly so that it is classified in the right category. The process description yielded 13 process steps; for each step the most important source of influence on the result of this process was given.

Number	Description	Most important source
1	Empty machine	operator
2	Look at planning list and select type and number	operator
3	Fetch cart with products	none
4	Get job number and stickers from bin	operator
5	Get job card	operator
6	Set up machine (if necessary)	operator/method
7	Throw products in bulk feeder	none
8	Switch machine on	none
9	[Machine measures and selects]	machine/material
10	Put bins on cart in selected order	operator
11	Announce batch to computer	none
12	Put batch in store	none
13	Announce position to computer	operator

The parts (activities) without influence on the result ('none') were ignored in the rest of the activity plan. Fetching the cart with products is such an example (at least in this process, with these carts, in this hall, *etc.*). But if, *e.g.*, the throughput time of this process was part of the assignment of the team, then this activity might well have been important. We will return to this PAT at the end of section 5.11, where some of its achievements are mentioned.

5·3 Cause and Effect Analysis (Step 2)

1 Summary

All important parts of the process (as established in step 1) are analysed in more detail. Everything which reasonably might cause a failure is summarized, and possible effects are mentioned.

2 Goal

All possible failure modes must be found out, and their effects should be indicated as clearly as possible. It is important that all team members are aware of these failure modes and effects.

3 *Result*

A list with causes and effects is established. Later on, this list can be used to define actions with the purpose of removing obstacles in the way to a controlled process. Thanks to the cross functionality of the team, the chances of obtaining a complete list are maximized. Another result is that the team members learn a lot about the background of how things are organized. In other words, knowledge is shared amongst all.

4 *Method*

All important process steps (from step 1 of the activity plan) are consequently analysed for possible failure modes. A useful tool is the cause and effect diagram (fish bone diagram) with its categories Man, Machine, Material, Method, and Environment. The chairman plays an important role as facilitator of the brainstorm process, he has to make sure that everyone participates. Therefore the use of brainstorm techniques is recommended. To report the result of this step it is convenient to use the numbering of step 1: the failure modes of part 1 are 1A, 1B, 1C, ..., those of part 2 are 2A, 2B, 2C, *etc.*.

5 *Guidelines for Use*

In this step we do not bother about the importance of the cause and effect combination, that will be done in the next step. Step 2 calls for real brainstorming to find all possible failure modes, with the emphasis on completeness.

The failure modes should not be described very profoundly in this phase. We restrict ourselves to the level of what operators observe. In the following step it is then possible to judge the frequency of the failures. If the failure mode is really important then in step 4 it will be necessary to dig deeper to find the root cause and to take corrective measures.

In the analysis of the effects we occasionally have to dig deeper, however, especially when inspections are dealt with. Direct inspection failures could be a wrong setting of a machine, a false judgement, *etc.*. But at a deeper level these failures might imply scrapped products or the breakdown of a machine. Feedback from internal or external customers is often valuable in this respect.

6 *Case Illustration*

All components of a surface mounted implosion diode (crystal, flange, stud, solder pellet) are assembled by hand with the use of jigs, and then they are enveloped by glass tubes. One jig contains a few thousand future diodes. Batches of six jigs are put in a furnace in which the products are imploded: the assemblies get heated under vacuum, the glass melts and becomes weak, and then under pressure flows around the components. After cooling down, the separate parts constitute a diode.

The process step of 'implosion' is sensitive to careless filling and rough treatment of the jigs. The amount of mechanical rejects is therefore rather high. But the furnace process is also critical, with the result that the electrical characteristics of the diodes can deteriorate badly. Enough reason to start a PAT for the jig fillers, who also operate the implosion furnace, in order to give them a tool to control the implosion process. The assignment of the PAT was limited at first to the implosion process itself.

Each of the 5 shifts delegated an operator to the PAT. Another operator was selected from measuring/selection machine, because as an internal customer he was confronted with the results of a bad implosion process. In their step 2 the team listed all causes of a bad implosion and the effects, part of which is shown in the next table. The causes are listed on the left and the effects on the right. Most effects are intermediate results, however, leading in the end to mechanical or electrical rejects. In the case illustration of the next step we continue with this PAT.

1 *Check furnace settings*	
1a Forgotten?	
1b Impossible?	possibility of wrong settings
1c Meters broken?	(and rejects/waste)
1d Incorrect readings?	
2 *Check diameters of graphite wheels and loosely tightened heat shields*	
2a No check	—furnace paddle may jam —paddle may go 'dancing' —quartz tube may break
3 *Remove products from furnace paddle*	
3a Not all products removed	—lower plate can not go down —quartz tube may break
4 *Put 6 implosion jigs on furnace paddle*	
4a Too rough	rejects
4b Paddle too high	higher chances of accidents and rejects
7 *Check implosion process*	
7a Time to vacuum too long	
□ at spooling	process is interrupted
□ before implosion	rejects
7b Power alarm	rejects

etcetera.

5·4 Risk Analysis (Step 3)

1 Summary

In this step the most important impediments to a controlled process are indicated. The importance of each cause and effect combination is determined by its inherent risk, a combination of the frequency of the cause, the severity of the effect, and the chance to prevent it happening. In fact, this step is some kind of process FMEA (Failure Mode and Effect Analysis, see also Stamatis, 1995).

2 Goal

The list of causes and effects may be very long, but in practice many combinations will be less relevant due to their infrequent occurrence or harmless effect. The risk analysis is meant to make objective difference between risky combinations and rather harmless ones. The goal is to obtain agreement among PAT members about the real important issues, because it often happens that different disciplines have their own ideas about what is important.

3 Result

After the risk analysis the list of causes and effects is ordered on the basis of their risk. This helps setting priorities about the causes of failures that should be attacked first.

4 Method

We start with the list of causes and effects of step 2 and work through the list completely. First the frequency of occurrence of a failure mode has to be established in a number between 1 ('hardly ever') and 10 ('always'). The following list might be useful:

Frequency

1	=	once per year (or hardly ever)
3	=	once per month (or sometimes)
5 or 6	=	once per week (or regularly)
8	=	once per shift (or often)
10	=	once per hour (or nearly always)

Intermediate numbers can be used for higher resolution (*e.g.*, 'once per day' might be a 7).

We then have to determine the severity of an effect. The result is again a number between 1 ('hardly any loss') and 10 ('failure at the customer'). The following list might be useful:

Severity

1	=	hardly any loss
3	=	small loss of time
5 or 6	=	repair, breakdown, loss of time at own work station
8	=	rejects, serious waste in own organization
10	=	rejects, failures at external customer

The final element defining the risk of a cause and effect combination is 'inspection'. Inspection influences the extend to which an effect will show up. So for each cause and effect combination we have to indicate how a failure is inspected. The success of this inspection is determined as a number from 1 ('very good') to 10 ('very bad'). This number, the detection, indicates how long it takes to find out that something went wrong. If a failure mode is detected (almost) immediately, and the effect is (almost) negligible, the detection is good, and a low value is appropriate. The following list might be used:

Detection

1	=	cause found almost immediately / effect hardly shows up
3	=	cause found in a few minutes
5 or 6	=	cause found in a quarter of an hour
8	=	cause found in an hour
10	=	cause found after hours / effect shows up completely

All three lists should be considered as guidelines. Each PAT has to decide its own criteria before starting the risk analysis.

The risk of a cause and effect combination is simply the product of the numbers 'frequency', 'severity' and 'detection':

$$Risk = Frequency \times Severity \times Detection$$

The risk is thus a number between 1 and 1000, and the higher the number the more risk is involved.

The scores of the team are filled out on the risk analysis sheet. The cause and effect combinations are listed in descending order of risk. The inspection is also included, together with the scores for frequency, severity and detection. The numbering of the previous steps is used again. An example of the risk analysis sheet is shown in the case illustration.

5 Guidelines for Use

The determination of frequency, severity, and detection can be rather involved. One reason is that many people have difficulties in considering the criteria independently. For instance, many people do not like to call an effect severe when the accompanying cause hardly ever occurs. And even more difficult is the case in which an oft occurring cause seldom leads to the severe effect, and instead mostly fizzles out. In this case we have two options: the first is to judge frequency and severity independently (worst case); the second is to distinguish the two possibilities and judge both alternatives separately. The first option, more then the second one, is aimed at preventing the cause, cost what it may. Both options have their merits, but it is important that the team selects one method and adheres to it throughout.

The detection is a judgement of how long it takes before the cause is detected, and to what extend the effect will show up. Whether the effect is severe and/or the cause occurs often is not important; that has been judged in the severity and frequency, respectively. It requires some abstract thinking to judge the detection correctly.

Also hard to fill out are the process steps that check something. These are of course inspections of other cause and effect combinations. Let us examine, *e.g.*, the check of the furnace settings (from the case illustration of the previous step), in more detail. This check is, in fact, an inspection of the failure mode 'wrong setting' with the effect 'rejects'. This check is so important that it has been made part of the regular process. The failure mode of this checking step is that it is not properly done, and the effect may be that the setting is not correct. But how can the check be inspected? Filling in a check sheet is a kind of inspection, but manipulation is still possible. The ultimate check is aimed at occurrence of the effect, but then it is too late, indeed! In short, in such a case the detection again judges the extent to which the effect occurs.

Every now and then there will be a discussion about the judgements. Process engineers have a tendency to estimate frequencies too low, while operators sometimes underestimate the severity. Monitoring frequencies for a while generally solves the first issue, and the second one can be solved by explanation of a customer or quality engineer.

6 Case Illustration

The cause and effect analysis of the PAT 'Implosion' is shown in the case illustration of step 2. In the following step the team judged the several combinations in accordance to the guidelines just discussed. A few operators in the PAT took sheets with them, and filled these out together with their colleagues in the shift. In three PAT meetings the separate sheets were combined into one; where scores differed strongly a discussion was held. Mostly the difference could be resolved, but in a few cases no scores were given yet, and actions were started to collect more data and gather more information.

In Table 5.1 part of the risk analysis sheet of the PAT is filled in. The list of cause and effect combinations is ordered on the basis of descending risk. Note that the team choose for the option to judge frequency and severity independently (combination 2A). Because, although graphite wheels and heat shields are often not checked, the listed effects—which are really serious—fortunately do not often occur.

The team judged the severity of effects to be heavier than we advised in our list. This is of course not a problem at all, because the risk analysis is just a tool for this team, without implications to others. A comparison between scores of

Table 5.1: Risk analysis sheet of the PAT 'Implosion'.

NUMBER	CAUSE	F	EFFECT	S	INSPECTION	D	RISK
2A	graphite wheels / heat shields not checked	8	—paddle may jam —paddle may go 'dancing' —quartz tube may break	10	visual (at drive in)	5	400
3A	products remain at paddle	5	—lower plate does not go down —quartz tube may break	10	visual (but hard to correct)	8	400
7B	power alarm of furnace	2	rejects	10	by process engineers	8	160
1B	check of furnace settings impossible	8	settings not correct	9	visual	2	144
1A	check of furnace settings forgotten	8	settings not correct	9	visual	2	108
1C	meters of furnace broken	3	settings not correct	6	visual	3	54
1D	meters of furnace incorrectly read	2	settings not correct	10	visual	2	40
4A	jigs too rough put on paddle	1	rejects	10	visual	3	30
4B	paddle too high	1	accidents and rejects	10	visual	1	20
7A.2	time to vacuum too long (before implosion)	2	rejects	10	alarm	1	20
7A.1	time to vacuum too long (at spooling)	3	process interrupted	3	alarm	1	9

this team and scores of other teams is neither necessary nor useful. The only goal of the risk analysis is that it provides a suitable order of topics for improvement for the members of this team.

In later steps the PAT has worked on all cause and effect combinations with a risk of 150 or higher. The approach was that the leader of the PAT, the responsible process engineer, defined a (small) project for every problem, with an operator as project leader. The process engineer kept himself available for advice. Thus a number of problems has been solved effectively by mostly simple and practical solutions stemming from the operators themselves. The problem with the graphite wheels (combination 2A), *e.g.*, has been solved by setting up a replacement scheme, and the introduction of a check sheet (to stress the importance of the check). In the project 'furnace paddle' (combination 3A) the operators did some research together with the engineer, and then they were able to introduce a new concept which limited the chance of spilled products drastically. Thus the process was made more efficient, and less maintenance of the furnace was called for.

We finally mention that all members of the PAT were taught some elementary knowledge of semiconductor physics. The reason behind this was that operators with some knowledge of the products can better judge the cause of rejects (bad crystals, wrong furnace settings, or other reasons). Together with this they were taught how to adjust the furnace settings. Thanks to these lessons the number of calls to the process engineers to check and adjust the furnace were reduced with 80%!

5·5 Improvement Actions (Step 4)

1 Summary

In steps 1–3 the bottlenecks of the process step have been listed and ordered. From this step on they will be resolved to control the process. In step 4 improvement actions will be defined and planned, and risks will be reduced.

2 Goal

To control a process step the impediments must be removed. From the previous steps the most important impediments are known, and are given a priority number. In this step the PAT seeks solutions for cause and effect combinations with high risks. If a solution is obvious an improvement action is started, and otherwise (when the team has yet no clue about possible solutions) the team decides how extra information can be obtained, perhaps new data must be collected and measurements be done.

3 Result

A list of improvement actions will be made up, with information about:

- □ The contents of the action;
- □ The responsible team member;
- □ The planning.

With this list the progress of the improvements can be checked.

4 Method

To find improvements the risk analysis sheet is a good starting point. We have to reduce the risk of a cause and effect combination, which can be done by improving at least one of the three scores (frequency, severity, detection). The biggest improvement can, of course, be made by reducing the highest score, but the severity is often difficult to influence. A severe effect should be overcome by preventing its cause from occurring (which is preferable) and/or a perfectly effective inspection.

In this step the PAT has to dig into the failure mode as deep as possible. A superficial solution is not acceptable, instead the root cause should be found and eliminated. Finding solutions is often a creative process which is given more chance through the cross functional character of the team. The chairman of the meetings has the task of guiding this process, perhaps through the use of brainstorming techniques and structured ways of working (such as the cause and effect diagram). He should also take care that the root cause is found; the motto is to dig deeper and deeper (ask 'why?' five times). The structured problem solving tools of Chapter 9 may also be useful.

For every improvement action the team selects the member who will perform it. Not all actions can be carried out by a member of the PAT, however; in that case the PAT appoints someone who is responsible for the deliberation with others and for the resulting planning.

5 Guidelines for Use

The normal task for process engineers and developers is to improve, therefore naturally they play an important role in this step. It would be unwise, however, not to utilise the capabilities of operators and others. We have had the experience that operators are very capable of finding solutions of practical problems, though help from engineers in bringing these solutions about is often necessary. Further advantages of operator involvement are that operators feel committed to the solutions (and give them a fair chance to be effective), that they better understand the process, and that they relieve the engineers of work.

Although step 4 is the formal place in the activity plan for improvement actions, actually from the very beginning improvements are possible. Listing them gives a nice overview of all activities of the team. During the meetings of the PAT this list can be discussed briefly, in order to monitor the progress. And the list is also a good means of communication with all other people involved.

To discuss the progress of the actions we advise removing the finished actions from this list, and putting them on a new list of finished actions with corresponding results. In this list of finished actions it is useful to include the new scores of frequency, severity, detection, and the resulting risk. One can then easily see whether or not the action was successful in reducing the risk. New actions are necessary when the risk is still too high.

6 Case Illustration

This case illustration describes the work of a Process Action Team from the factory where leaded glass diodes are assembled, and which team had the assignment to control the measuring/selection process. The need to measure and select diodes has already been discussed in the case illustration of step 1. The name of the PAT ('LGDynamic') was chosen for three reasons: (1) indicating the name of the product (LGD); (2) showing its energetic character; (3) referring to the most difficult measurements, the 'dynamic' measurements.

From the fourth meeting onwards, when the risk analysis sheet was filled in (and some minor corrections of the cause and effect analysis were made), improvement actions were defined. Altogether the PAT has started nearly 50 actions about all kinds of issues: procedures, organization, machine adjustments, work instructions, *etc.*. In Table 5.2 the action list of the meeting of 30th January is shown. Note that the finished actions have been removed from the list:

The final result of this PAT is that the process has many fewer disturbances, thanks to several improvements of the machines (to which action 19 contributed). The adjustments to switch from one type to another are now much simpler (thanks to a.o. actions 16 and 20), which yielded fewer mechanical rejects. To control the process of selection of diodes the team introduced a system with 'gauge diodes' (action 23); diodes with known characteristics are measured regularly, and when the actual measurements differ too much from the known values an out of control action plan comes into operation. Thanks to this scheme all operators (and not

Table 5.2: Action List of the PAT 'LGDynamics' (30/1).

No.	ACTION	WHO	WHEN	STATUS
16	2-state switch for air pressure (machines 018 and 019) [18 C/D/E]	EBN, RSS	a.s.a.p.	ordered; expected around 20/2
17	examine 'good' products rejected by the sample handler [16A]	operators EBN	until further notice	no such products last week
18	fill in log book for mechanical rejects [18 C/D/E]	operators	weeks 4–6	introduced
19	adjust tray 1 of machine 18 again [18 C/D/E]	EBN	23/1	when planning permits (perhaps on 2/2)
20	contrive simpler adjustments [18 C/D/E]	RSS	30/1	chain and air; simpler magnets; new concepts
23	summarise pros and cons of gauge diodes [16–18]	RKG, EKP, EBN	30/1	brainstorm session held
24	organize study journey to MeCoVe–line	GHT	6/2	
25	make list with normal and deviating yields [17B]	EBN	27/2	
26	improve R_d measurements; announce in technology meeting [18F]	EBN	6/2	

just the experienced ones) can quickly find the cause of the disturbance and take corrective actions.

5·6 Measurements (Step 5)

1 Summary

In this step the necessary measurements are determined, and how they should be done, for how long, by whom, *etc.*.

2 Goal

Extra information is required when there is not yet an obvious solution to a problem. The PAT decides which measurements are needed for this purpose, thus making sure that reliable data is collected. In this step the basis of the control chart is provided also.

3 Result

The team agrees about measurements to be done. For each measurement the following subjects are described:

- ☐ The reason for the measurement;

- ☐ The measurement itself;
- ☐ The measurement procedure and equipment;
- ☐ The frequency;
- ☐ The observer;
- ☐ The period of collecting data;
- ☐ The responsible team member.

When enough data is collected the team does the analysis. Nearly always the '7 tools' are sufficient. If an obvious solution to a problem can still not be found, the team has to decide whether it wants to continue this measurements, or whether a new one might be more appropriate.

The result of collecting data is that the team will have sufficient information to decide on an improvement action, and that it has a firm basis for the introduction of a control chart (step 7 of the activity plan).

4 *Method*

If a PAT is not yet able to find improvements for a high risk cause and effect combination, then more information is needed. We might study the effect for a while in order to examine the variation (are there differences between shifts? between batches? within batches? within products? between machines? ...). Another possibility is to focus on causes. The variation of a chemical bath might be examined by measuring the concentrations and temperature regularly. Or we could record the periods between successive interventions in the composition of the bath, and the amounts of added chemicals.

The data can often be analysed efficiently with relatively simple tools such as the trend graph, histogram, or scatter diagram (see section 7.3). This analysis will often be sufficient for an improvement action; only occasionally will the expertise of a statistician be required for a more thorough study.

5 *Guidelines for Use*

The pithy American saying 'No data, no talk' is a strong reminder of that feelings can be converted into facts through measurements. Also, in a PAT differences of opinion will occur, and an important way of overcome squabbling is to test opinions with data. When the way of working of two shifts is somewhat different we might try to examine the effects quantitatively. When the limits of process settings are not clear the process engineer has to support them with data, and so on.

Again, it is true that operators are important in deciding upon measurements. They can make or mar the value of the data, because in the end they are often the ones who have to do the work. Their input may lead to simple and feasible arrangements, with the result that the collected data is 100% reliable—an essential prerequisite if one wants to draw conclusions.

6 *Case Illustration*

A stack diode is a high voltage diode obtained by stacking a number of crystals. A stack diode is therefore longer than other diodes. To passivate the diode (to isolate the crystal from the outside world) 2–4 beads of glass slurry are deposited around the crystals, and then the glass is sintered in a furnace. Both process steps, glass deposition, and sintering, are critical in the quality. Thus the SPC steering committee decided to install PATs with the assignment of bringing these process steps under control.

PAT 'SINT' was responsible for the sintering process and concentrated on the biggest problem: mechanical rejects. The team started collecting data on four categories of mechanical rejects:

- 'Spiders' (some clotted diodes);
- 'Holes' (in the glass);
- 'Flow backs' (of the glass);
- 'Ugly forms' (deviations from the ideal cigar form).

During a six week period the operator of the night shift counted the rejects of eight furnace loads (one from each furnace tube). The position in the tube was also recorded. In this period nine different types of stack diodes were produced. The analysis of the detailed reject data revealed that:

- □ The longer the diode, the more rejects;
- □ Occasionally the number of 'flow backs' is large, but only for some types;
- □ Rejects of the 'spider' type occur regularly;
- □ Nearly half of the 'spiders' come from the second furnace tube;
- □ The number of 'holes' is small.

Based on this analysis new measurements were introduced, with the focus on 'spiders' (which happen to be typical sintering problems). Especially was the difference between furnace tubes interesting. For several months the alarm signals of the eight tubes were counted (by the operator responsible for the daily management of the furnace). Fortunately the furnace software keeps track of the alarms, so continuous monitoring was not necessary. During this period the same operator also registered the temperature and pressure profiles of the tubes on a monthly basis. Thanks to these measurements the tubes could be mutually compared, and the reason why the second tube had so many more 'spiders' was able to be found. With the combined information the PAT was able to determine the critical parts of the sintering process, and set up an out of control action plan (step 8 of the activity plan).

5·7 Measurement Analysis (Step 6)

1 Summary

This is the first step in monitoring the process step. Since we rely heavily on measurements and statistics we need to know whether the data can be trusted. A measurement can be biased (that is the reason for calibrating the system regularly), but it also adds a certain amount of variation to the results. In this step a small experiment is carried out in order to investigate the magnitude of this measurement variation.

2 Goal

Some products are measured repeatedly by different operators according to a standard scheme. This allows us to determinate the precision of the measurement. If it does not satisfy a generally accepted standard (see the manual of Chrysler, Ford, and GM, 1990) then the measurement procedure needs improvement. This is not always possible, however, and if it is not then a solution has to be found for the problem of individual measurements being able to be affected with unacceptably large measurement errors.

3 Result

After this step the measurement variation owed to the equipment itself (the so called 'repeatability') is quantified, as well as the extra variation owed to the procedure (the 'reproducibility'). The analysis is therefore also known as an R&R study. The total measurement variation is the combination of the two, and th standard is that it should be less than 30% of the tolerance width. The followi situation can be distinguished if this standard is not met:

- Repeatability is the most important source of measurement variation equipment is not suitable for the application;
- Reproducibility is the most important source of measurement v The procedure has to be improved, with the purpose of reducing ence between operators.

If repeatability variation cannot be reduced to an acceptable level, th of measurement variation has to be reduced by repeated measuremen averages. But a new measurement procedure is needed if reproduc be improved sufficiently (perhaps combined with better equipment)

4 Method

For *non-destructive* measurements (products can be measured rep layout of Table 5.3 can be used to evaluate the measurement preci 10 products from the total range of the process variation, and select Each operator measures each product three times in a row, and write down on a sheet. The calculation method then tells step by step (with different numbers of operators and/or repetitions the calculations have to be adapted with different constants; see Table 5.4).

Incidentally, the ***resolution*** of the measurement should be sufficient as well; as a rule of thumb it should be possible to divide the range of the process's variation into at least 10 measurement values.

Table 5.3: Layout for a measurement analysis. 'Av.' and $\overline{X}$ are abbreviations of the average and R for the range (the difference between maximum and minimum).

Pro-duct	Operator 1				Operator 2				Operator 3			
	1	2	3	R	1	2	3	R	1	2	3	R
1												
2												
3												
4												
5												
6												
7												
8												
9												
10												
Av.												
				$\overline{R}_1$				$\overline{R}_2$				$\overline{R}_3$
Av.	$\overline{X}_1$				$\overline{X}_2$				$\overline{X}_3$			

Outliers

Average range: $\overline{R} = (\overline{R}_1 + \overline{R}_2 + \overline{R}_3)/3$.

Ranges higher than $D_4 \times \overline{R}$ need to be examined carefully, and perhaps the orresponding measurements should be corrected or measured again ($D_4 = 2.58$ ith 3 repetitions).

peatability and Reproducibility

eatability (equipment variation) $= K_1 \times \overline{R}$. ($K_1 = 3.05$ with 3 repetitions).
roducibility (operator variation) $= K_2 \times X_d$ with X_d the range of $\overline{X}_1$, $\overline{X}_2$ and
$K_2 = 2.70$ with 3 operators).

Total Measurement Variation and %–Tolerance

Total measurement variation = $\sqrt{(\text{repeatability})^2 + (\text{reproducibility})^2}$.

The total measurement variation is compared with the tolerance width, and is expressed as a percentage:

%–tolerance = [total measurement variation / tolerance width] × 100.

Judgement

The total measurement variation is judged by the %–tolerance. The criterion is:

$< 10\%$	acceptable measurement precision;
10%–30%	perhaps acceptable measurement precision (depends upon the application);
$> 30\%$	unacceptable measurement precision.

Table 5.4: Constants for the calculation of repeatability and reproducibility (provided that the product of the number of operators with the number of products is more than 10).

Repetitions	D_4	K_1	Operators	K_2
2	3.27	4.56	2	3.65
3	2.58	3.05	3	2.70
4	2.28	2.50	4	2.30
5	2.12	2.21	5	2.08

5 Guidelines for Use

To avoid operators being influenced by the results of their colleagues, it is better to ensure they use their own sheet. Afterwards the results can be filled in on one common sheet. We advise using at least as many operators as there are shifts, then all the shifts are involved. If there are more than three shifts then the measurement variation will be estimated even better. Another remark about the performance of the experiment is that the measurements are preferably done in a random order. As with any experiment any unknown systematic effects can thus be avoided as well as possible.

The scatter plot of the measurement value against the product is a useful graph for obtaining a first impression of the measurement variation. In Figure 5.3 such a graph is shown, with the thickness of a product on the vertical axis, the product itself on the horizontal axis, and the number of the operator plotted in the graph. The figure is associated with an analysis of the thickness measurements of products, with four operators involved (see the case illustration).

The percentage tolerance cannot be calculated if there is only one limit. In this situation the measurement variation is usually compared with the process

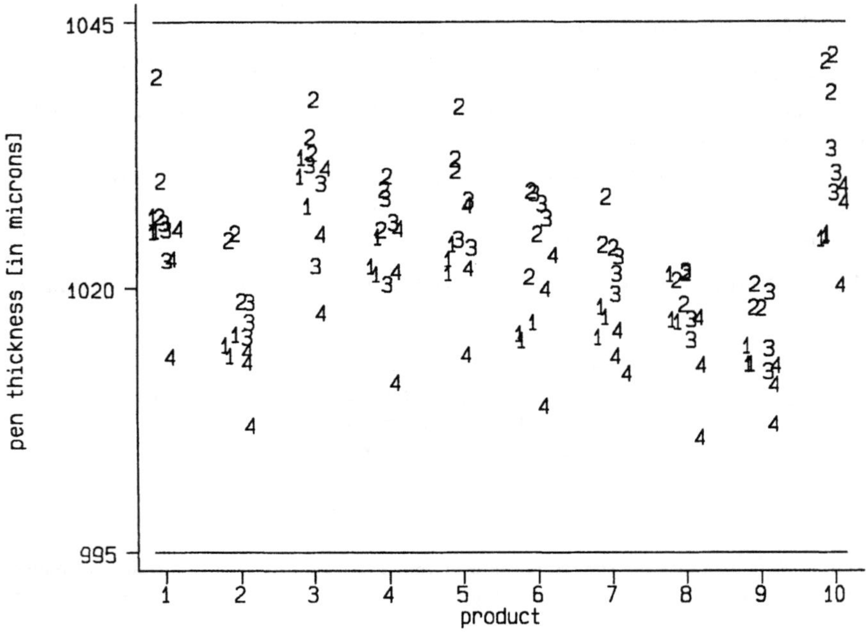

Figure 5.3: Graph of the measurement analysis of measuring thicknesses.

variation. The requirement for an acceptable measurement method is that the total variation is at most 30% of the process's variation. Incidentally, this same requirement can also be of use when there are two specification limits; and often it is used in this case as well.

A measurement analysis is still possible when the measurement is not a variable but an attribute (see the Chrysler, Ford and GM manual). The method amounts to judging some products several times by a few observers. A condition is that all products have to be classified objectively into one of two classes good or false. The requirement for a precise measurement method is that both within and between observers a high degree of agreement exists. We have had the experience that such experiments are very difficult to design, because experienced observers are very good at recognizing products they have seen before. Thus tricks are needed to 'deceive' the experienced observer. Another disadvantage of this measurement analysis is that the result is nearly always unacceptable. But that may as well be the message of the method: an objective measurement is clearly preferable to a subjective judgement.

Engel and De Vries (1997) is a useful reference for the criteria, as well as to other literature about the measurement analysis.

6 *Case Illustration*

The thickness of pens in a glass base (the connector with the electron gun in a picture tube or monitor) is critical. Thick pens do not fit, and thin pens break easily. In the final process step of glass bases the thickness of these pens is determined in a pickling process.

All process operators participated in a PAT to bring the pickling process under control. Since the baths are large they are quite stable, so the chemical composition of the baths is not the main concern. The temperature is much more variable, with a large effect on the thickness. For this reason a sample from each rack of glass bases is selected for measurements of pen thickness. In the case of large deviations from the target the process has to be adjusted by increasing or reducing the pickling time.

The thickness is measured with a digital micrometer. The quality of the measurement method is examined with the measurement analysis of step 6. Each one of the four operators carried out his share of measurements, measuring 10 products three times in a row, in his normal way. The data is shown in Table 5.5 (all numbers should have 1000 μm added to them). The thickness is specified to be within 995 and 1045 μm, so the tolerance width is 50 μm. In Figure 5.3 the data is shown in a scatter plot, with the number of the operator plotted.

Table 5.5: Data of the measurement analysis of pen thickness (all data should be have 1000 μm added to them).

Pro-	Operator 1				Operator 2				Operator 3				Operator 4			
duct	1	2	3	R	1	2	3	R	1	2	3	R	1	2	3	R
1	24	25	26	2	26	30	39	13	22	25	25	3	13	22	25	12
2	13	14	15	2	18	24	24	6	15	16	18	3	6	12	13	7
3	27	30	31	4	32	34	37	5	21	29	31	10	17	25	30	13
4	21	21	24	3	25	28	30	5	20	26	28	8	10	21	25	15
5	21	22	23	2	30	32	36	6	23	24	27	4	13	21	27	14
6	14	15	16	2	20	24	28	8	26	27	28	2	8	19	22	14
7	15	17	17	2	23	23	28	5	19	21	22	3	11	13	15	4
8	16	16	21	5	18	20	21	3	14	16	21	7	5	12	17	12
9	12	12	14	2	17	17	20	3	11	14	19	8	6	10	12	6
10	24	24	24	0	38	40	41	3	28	30	32	4	20	27	29	9
Av.				2.4				5.7				5.2				10.6
				$\overline{R}_1$				$\overline{R}_2$				$\overline{R}_3$				$\overline{R}_4$
Av.	$\overline{X}_1$	19.80			$\overline{X}_2$	27.43			$\overline{X}_3$	22.60			$\overline{X}_4$	16.87		

Outliers

Average range: $\overline{R} = (2.4 + 5.7 + 5.2 + 10.6)/4 = 5.975$.

Ranges above $D_4 \times \overline{R} = 2.58 \times 5.975 = 15.4$ are suspect; maximum range in the table is 15, however. Consequently there are no outliers.

Repeatability and Reproducibility

Repeatability (equipment variation) = $K_1 \times \overline{R} = 3.05 * 5.975 = 18.22$.
Reproducibility (operator variation) = $K_2 \times X_d = 2.30 * 10.57 = 24.30$, ($X_d = 27.43 - 16.87 = 10.57$; $K_2 = 2.30$ with 4 operators).

Total Measurement Variation and %-Tolerance

Total measurement variation = $\sqrt{(\text{repeatability})^2 + (\text{reproducibility})^2}$

$= \sqrt{(18.22^2 + 24.30^2)} = \sqrt{(331.97 + 590.49)} = \sqrt{922.46} = 30.37$.

%-tolerance = [total measurement variation/tolerance width] × 100

$= 30.37/50 = 61\%$.

The conclusion was that the measurement method was completely inadequate for this application. Operator variation was the most important source of variation. On closer examination it turned out that the operators measured the pens at different locations, an important fact because in general pens turn out to be smaller at the end. The operators then agreed to measure as well as possible in the middle of the pen. This improved reproducibility considerably.

When we study Table 5.5 more closely we can see some systematisation in the data. The three measurements in a row are nearly always in ascending order, and the reason turned out to be a problem with the micrometer. Adjusting the equipment regularly reduced the equipment's variation. The combination of these two adaptations of the procedure improved the measurement method to an acceptable 12% tolerance.

5·8 Control Chart (Step 7)

1 *Summary*

SPC is often equated with control charts. But only in this step of the activity plan now bottlenecks have been identified and removed, and it is known that the measurements to be used in the chart are reliable, do the control charts start to play their part. In step 7 the measurements are analysed in more detail and one or more control charts are constructed. This step applies to all measurements if one is not sufficient.

2 *Goal*

This step leads to a better insight in the most important measurement(s) for process monitoring. And it also leads to the introduction of control charts for one or more characteristics of the measurements (*e.g.*, average and range). With control charts we can easily decide whether the process is still under control or out of control.

3 Result

After this step the measurement is clearly described. Based on recent data it is known:

- □ Which characteristics are important (average, range, other aspects?);
- □ Whether these characteristics are under control;
- □ What limits belong to the process inherent variation.

For every characteristic the PAT designed a control chart by which the characteristic will be monitored.

4 Method

At least 20–25 samples of the measurement are needed before we can state with sufficient confidence the status of the process and the limits of the process' inherent variation.

First of all we need to decide upon the characteristics to be analysed. The common summary of a sample into the mean and the spread, leading to the $\overline{X}$-R chart (or $\overline{X}$-s chart), too often gives a deceptive form of process monitoring. The correct summary is determined mainly by the structure of the sample and the nature of the process. In the case illustration we show how a sample of four measurements is summarized into three useful characteristics. Knowledge of the process and the sources of variation, in combination with a good process diagnosis, leads to the correct characteristics. Chapters 7 and 8 will deal with this problem in more detail.

A trend graph of a characteristic shows its behaviour in time: outliers, jumps, trends, and patterns will become visible. In all these situations the characteristic is not under control because it is not 'constant' in time. Apart from the common causes of variation one or more special causes are active. Not many characteristics are under control, however, if this strict definition of being under control is used. Therefore, when judging whether or not a process is under control we tend to be flexible: a few outliers or 'small' jumps, trends or patterns are allowed for a controlled process.

Now only the operational definition of the limits of the process inherent variation lacks, with the meaning that a value beyond these limits indicates the occurrence of a special cause of variation. A standard recipe for the calculation of these limits (also known as '3σ limits') is the 'moving range' method:

- □ The successive values of the characteristic are:

$$X_1, X_2, \ldots, X_n;$$

- □ The mean value is $\overline{X}$;
- □ The successive ranges ('moving ranges') of two values are:

$$MR_1 = |X_1 - X_2|, \quad MR_2 = |X_2 - X_3|, \quad \ldots, \quad MR_{n-1} = |X_{n-1} - Z_n|;$$

- The mean moving range is $\overline{MR}$;
- The limits of the process' inherent variation are:

$$\overline{X} \pm 2.66 \times \overline{MR}.$$

Other methods of calculating the limits of the process inherent variation are described in Chapters 7 and 8.

The ingredients of the control chart are ready now. The upper limit of the process' inherent variation is the Upper Control Limit (UCL), and the lower limit is the Lower Control Limit (LCL). The third line in the control chart is the Central Line (CL), which is either the mean value or the target value. After each sample the value of the characteristic is entered in the control chart.

We finally remark that the control limits are often determined in two steps. We start with the complete data set, and out of control values are determined. Then new limits are calculated with these out of control values excluded, because these measurements obviously do not belong to the process inherent variation.

5 *Guidelines for Use*

The control chart is the best known part of SPC, and in the history of industrial statistics much has already been investigated (see Chapters 6–8). With the control charts the process is monitored. If a characteristic exceeds a control line, then the deviation is larger than normal and a search for a special cause is useful (the out of control action plan of the next step should give guidance in this search). After elimination of this cause the process is back under control. But apart from this signal the chart may show up other evidence of the process being out of control: jumps, trends, patterns. These may be hard to detect if the control chart signals only points beyond the control limits. That is why over the years many additional signalling rules have been invented, such as 'six points in a row descending or ascending', or 'nine consecutive points above or below the central line'. It is useful to extend the use of the control chart with some of these rules; see also Chapter 6.

6 *Case Illustration*

One of the parts of a diode is the stud (see the diode of Figure 5.2). The studs of some types of diodes need a silver layer to decrease the melting point at soldering the lead–stud junction sufficiently. Too thick a layer, however, causes some electrical characteristics to deteriorate unacceptably. The silver layer is put on the studs in an evaporation process, in batches of 75 jigs simultaneously. The layer thickness is monitored for every batch by measuring the thickness on two copper slices, one at the bottom of the system and one at the top. The assignment of a PAT was to bring this process under control.

After the preliminary steps the team designed and implemented control charts of the thickness of the silver layer. In the measurement analysis (step 6) the precision turned out to be insufficient (40% of the tolerance width was used), so the team decided to measure the copper slices twice, in order to diminish the effect of measurement errors.

The analysis of past data showed that the common $\overline{X}$-R chart did not give a correct representation of the process variation. The variation between slices turned out to be much larger than the variation owed to repeated measurements. Moreover, a big difference between the top slice and the bottom slice is interesting. In the $\overline{X}$-R chart both effects (variation in repeated measurements, and systematic differences between top and bottom) are combined in the R-chart.

For monitoring layer thickness a better control chart system was needed. The PAT defined three characteristics to be examined by a chart:

- □ The mean layer thickness;
- □ The difference between top and bottom;
- □ The mean range of the repeated measurements.

All characteristics have a clear meaning, and dedicated out of control action plans could be devised for corrective actions. The mean thickness is an indication of the amount of evaporated silver; the difference between top and bottom is an indication of the variation within the evaporation system; and the mean range of the repeated measurements in an indication of the reliability of the measurement equipment. In Figures 5.4–5.6 the control charts of the three characteristics are shown (for only a part of the total period).

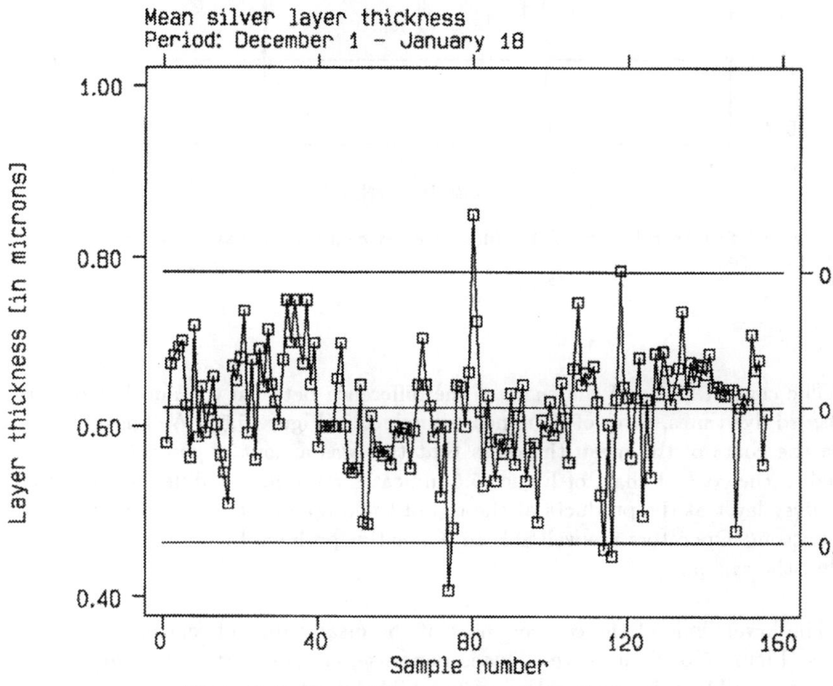

Figure 5.4: Control chart of mean layer thickness.

The graph of the mean layer of the 156 samples (Figure 5.4) makes clear that there are small jumps: some periods the mean is above (near samples 35 and 140), and some periods below (near sample 60) the central line. The control limits were exceeded in some turbulent periods (near samples 80 and 120). The control chart tells us therefore that changing circumstances influence the amount of evaporated silver, and that real out of control situations occur every now and then.

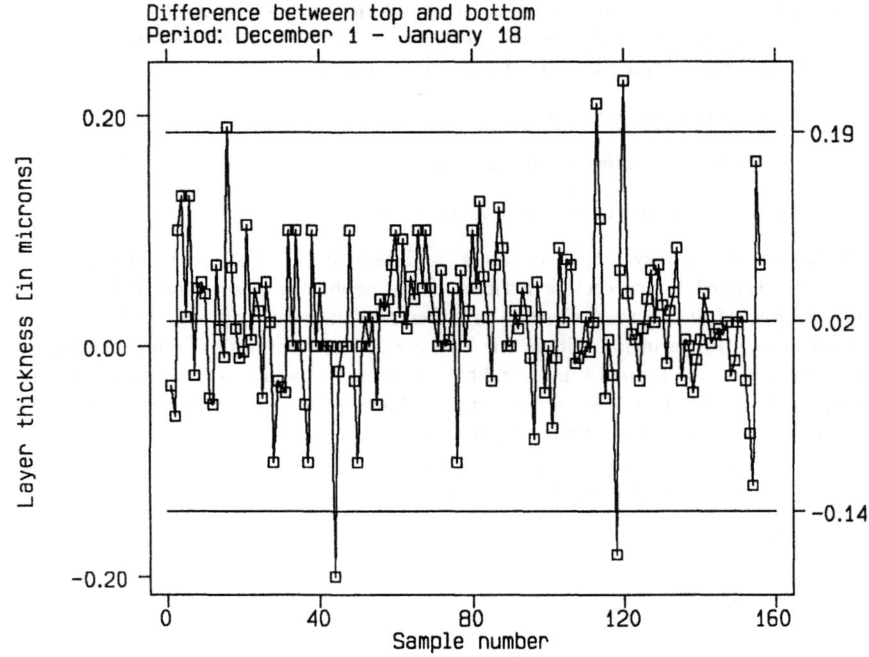

Figure 5.5: Control chart of the difference between top slice and bottom slice.

The control limits of the chart of the difference between top and bottom are exceeded five times, three of them near sample 120 (Figure 5.5). We know already from the chart of the mean thickness that this was an out of control period. A signal in the control chart of Figure 5.5 indicates a significant difference between the silver layer at the products at the top of the evaporation system and those at the bottom. Operators are well able to repair this problem by adjusting reflectors within the system.

The lower limit of the control chart of the mean range of repeated measurements (Figure 5.6) is negative. This is nonsense, of course, since the range itself can not possibly take on negative values. The chart is therefore not exactly a beauty. Nonetheless it shows that in the period from samples 30–90 a special pattern occurred: many ranges were either 0 or 0.05. It turned out that the measuring equipment had faltered: normally the thickness can be read to two significant decimals, but in that period only one digit was significant.

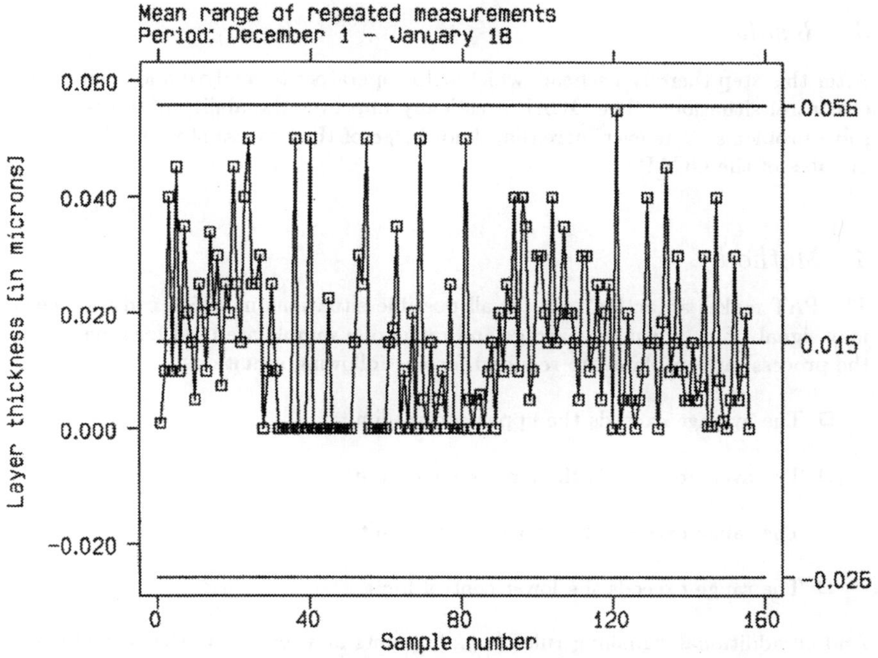

Figure 5.6: Control chart of the mean range of repeated measurements.

5·9 Out of Control Action Plan (Step 8)

1 *Summary*

The value of a control chart is only limited in practice if operators have no means of correcting an out of control situation. That is why in step 8 the PAT produces a flow chart (or description) of all checks and activities to be carried out when things go wrong. The out of control action plan may be considered as the most practical description of knowledge and experience of the process step.

2 *Goal*

In an out of control process the cause of the disturbance has to be found as quickly as possible. The out of control action plan (or OCAP, see Sandorf and Bassett, 1993) meets this aim. All available diagnostic knowledge is inserted, and all operators can solve a problem in the same way. The operators should be given maximal opportunities to solve the problems themselves; mechanics and process engineers have to describe their solutions as well as possible, and delegate their activities to the people immediately involved in the process: the operators.

3 Result

After this step there is a scheme which helps operators to analyse and correct out of control situations. The OCAP drastically improves the ability of operators to solve problems. The ever increasing knowledge of the process step is reflected in updates of the OCAP.

4 Method

The PAT makes an action plan for all possible situations in which control charts may signal. *E.g.*, when the average and range of a sample are used for monitoring the process, action plans are required for the following situations:

- □ The average exceeds the upper control limit;
- □ The average exceeds the lower control limit;
- □ The range exceeds the upper control limit;
- □ The range exceeds the lower control limit.

And all additional signalling rules (*e.g.*, '9 points in a row above the central line') require their own OCAPs.

The OCAP refers to two elements: process and product. The control chart indicates that the *process* is out of control; this has to be remedied, of course. By following a path with one or more checks, the user of the OCAP will be guided towards an action to manage this situation. But the *products* from this out of control process may also be different from the 'normal' products; the OCAP should also serve as a guide for what to do with these products.

It is not easy to make a good OCAP. The logical sequence of checks and actions should be determined, and the result may be an extended scheme. Therefore it has to be presented in a simple form when it is meant as a practical tool for operators. Operators should be endeavoured to solve the problems themselves, without calling maintenance mechanics or process engineers. A better OCAP can be obtained if the original solution 'call the process engineer' can be replaced with the analysis and/or solution of this engineer or mechanic, in such a way that the operators can do this themselves. Of course, this is not always possible, but organisations in which operators are both considered important and are effective can go a very long way down this track. Operators may need specific training and/or other help, but the yield: quicker solutions and less time needed from more expensive employees, will generally quickly outbalance the cost.

5 Guidelines for Use

Out of Control Action Plans and control charts are two sides of the same coin; we even dare to say that a control chart without OCAP is of limited value. Even without control charts, but with visual checks, for example, OCAPs can be effective tools for guiding operators towards solutions.

6 Case Illustration

The implosion process (when assembling diodes) was discussed in the case illustration of step 2. Here we show the out of control action plan of the implosion process of the surface mount glass diode. The PAT which made this OCAP had the assignment of bringing the process under control, and of setting up checks which would safeguard customers from faulty products.

The PAT decided to introduce two measurements/checks per day on every furnace tube:

- □ The glass test;
- □ A visual check.

The glass test is a measurement of the inner diameter of four glass tubes after the implosion process. For the other test a sample of 50 products is checked for several failures, each type of which is illustrated by a photograph above the work station. The failures can be attributed to a specific cause which the OCAP helps in removing. As an example, glass at the bottom of the flange means that the product was too hot during implosion. One possible cause is that the wrong kind of glass has been used, but another cause is that the temperature in the furnace tube was too high. In the latter case the temperature needs adjustment.

The four measurements of the inner diameter, obtained from the glass test, are summarized in an $\overline{X}$-R chart. The corresponding OCAP is the following question–answer scheme.

Is the average within the control limits?

1 YES: Is the range within the control limits?

11 YES: *Continue production*

12 NO: Is this true for more than one furnace tubes?

121 YES: *Throw products away*
Empty glass filling machine
Check other furnace tubes
Take new glass

122 NO: *Put products aside and make note*
Put furnace tube out of use

2 NO: Was a wrong program used?

21 YES: *Throw products away*
Check jigs
Set correct program
Repeat glass test afterwards

22 NO: Has the program been interrupted?

221 YES: *Put products aside and make note*
Repeat glass test afterwards

222 NO: Have glass types been interchanged or mixed up?

222.1 YES: *Action 121*

222.2 NO: Is the setting of the process gas wrong?

222.21 YES: *Put products aside for process engineer*
Set correct N_2 flow
Repeat glass test afterwards

222.22 NO: Does pressure bullet stick?

222.221 YES: *Put products aside for process engineer*
Clean pressure bullet
Repeat glass test afterwards

222.222 NO: Are there problems with the paddle?

222.222.1 YES: *Put products aside for process engineer*
Check furnace tube mechanics
Has there been a current interruption?
Repeat glass test afterwards

222.222.2 NO: **STOP PRODUCTION;**
CALL PROCESS ENGINEER

Some readers might prefer the flow chart of Figure 5.7 (with a more concise description of the actions). Note that the actions refer to a correction of the process and to the status of the products. Sometimes the products have to be thrown away immediately, in other situations they are put aside for further judgement (from the quality assurance engineer or the process engineer). The OCAP reflects how operators deal with a problem. They start checking their own responsibility (wrong program, mixed glass), and then the outside causes. Owing to process improvements the order of the most probable causes is likely to change. That is why this OCAP will be updated regularly.

5·10 Process Capability Study (Step 9)

1 Summary

In this step the status of the process step is described: is the process under control? and does it meet the specifications?

2 Goal

The process capability study describes: (1) whether or not the team succeeded in bringing the process step under control; (2) whether or not the specifications are met. The process capability study is used by the process' owner to decide whether

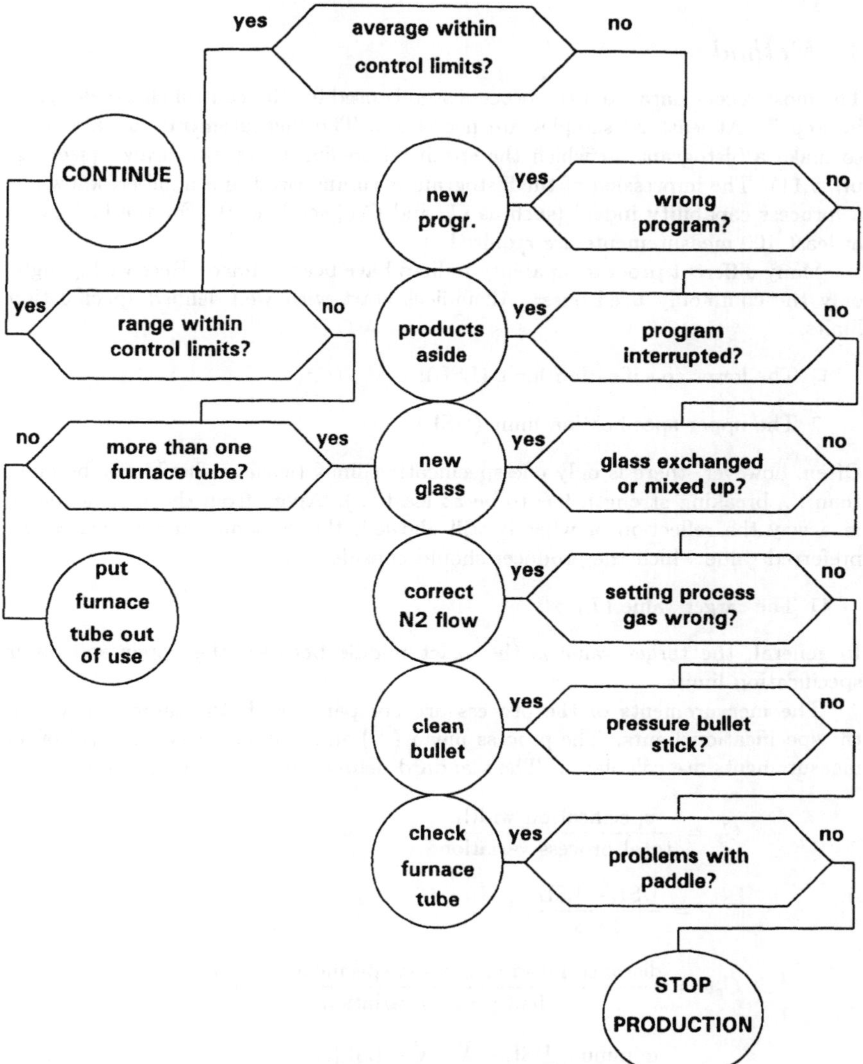

Figure 5.7: OCAP of the glass test.

extra activities are needed to improve the process step substantially (a breakthrough). If the process performs well enough, then the normal maintenance activities of step 10 will be sufficient.

3 Result

The process capability study is as concise as possible a summary of all management information necessary to control the process step. The selected characteristics (from step 7) for monitoring the process are all included with a control chart, and a histogram of the most recent data shows the conformance of the process with the specifications. A process capability index (*e.g.*, C_{pk}) quantifies this conformance.

4 *Method*

The most recent data from the process step is used for the control charts designed in step 7. At least 20 samples are necessary. The measurements are also used to make a histogram, in which the specification limits are also drawn (see Figure 5.11). The impression of the histogram is summarized in a number, known as a 'process capability index' (such as C_p and C_{pk}, see below). For the histogram at least 100 measurements are required.

Many different process capability indices have been defined. Here we highlight only the commonly used ones. All indices start with well defined specification limits:

- □ The lower specification limit (LSL);
- □ The upper specification limit (USL).

Often, however, there is only one specification limit (leakage should not be more than ..., breaking strength has to be at least ...). Apart from the specifications, in a way the reflection of what is still allowed, the customer also indicates the preferred value which the producer should consider as:

- □ The target value (T).

In general, the target value is the exact middle between the upper and lower specification limits.

The measurements of the process are compared with the target value and the specification limits. The process mean ($\overline{X}$) and standard deviation (s) of all measurements are calculated. The standard definitions of C_p and C_{pk} are:

$$\begin{aligned} C_p &= \frac{\text{specification width}}{\text{total process variation}} \\ &= \frac{\text{USL} - \text{LSL}}{6 \times s}; \\ C_{pk} &= \frac{\text{distance mean to nearest specification limit}}{\text{half process variation}} \\ &= \frac{\text{minimum}[\text{USL} - \overline{X},\ \overline{X} - \text{LSL}]}{3 \times s}. \end{aligned} \tag{5.1}$$

C_{pk} is set to 0 is the calculation yields a negative number.

In a one-sided situation the index C_p is not defined, and the definition of C_{pk} depends on having a target value or not. Assume that there is only a lower specification limit, and no target value. Then:

$$C_{pk} = \frac{\overline{X} - \text{LSL}}{3 \times s}.$$

If there is a target value then the above formula has to be corrected for a possible deviation:

$$C_{pk} = \frac{(T - \text{LSL}) - |\overline{X} - T|}{3 \times s}.$$

(note that both C_{pk} values coincide only if $\overline{X} \leqslant T$). The adaptations for the situation with only an upper limit are obvious.

Common adaptations of C_p and C_{pk}, when the target value is not exactly the average of lower and upper specification limits, are:

$$C_p = \frac{\text{minimum}[\text{USL} - T,\ T - \text{LSL}]}{3 \times s},$$

$$C_{pk} = \frac{\text{minimum}[\text{USL} - T,\ T - \text{LSL}] - |\overline{X} - T|}{3 \times s}.$$

5 Guidelines for Use

C_p compares the tolerance width and the process variation, irrespective of the location of the process. C_{pk} does take the location into account and is always smaller than C_p, unless the process average is equal to the target value. The demands on processes are continuously increasing, which is reflected in higher and higher values of C_{pk}: nowadays it not uncommon for values of at least 1.67 to be demanded!

Calculating C_p and C_{pk} is not difficult, interpretation can be very hard, however.

1 C_p and C_{pk} have no predictive value at all when the process is not under control.

2 The relation of C_p and C_{pk} with a percentage reject should be based on the underlying probability distribution. If this is the normal distribution, then $C_{pk} = 1$ means at most 0.27% rejects. But $C_{pk} = 1$ might mean any percentage rejects between 0% and over 10% if the distribution is not known.

3 C_p and C_{pk} are just estimates, based on a sample. If the sample is small or biased, then the estimate may deviate considerably from reality (the population value).

4 The meaning of C_p or C_{pk} is not clear when the distribution of the measurements is skewed (not symmetrical). A transformation of the data (*e.g.*, the logarithm, or the square root) might make the distribution more symmetrical, and the meaning of a consequently calculated capability index better understood.

The interested reader is referred to Kane (1986), Gunter (1989), or Bissel (1990).

6 Case Illustration

Diodes—and other electrical components—are soldered on printed circuit boards, so a small amount of tin is needed on the leads or flanges of the diode. Tin plating is therefore an important process step in assembling diodes. Leaded implosion diodes receive a tin layer in a two step process: first the diodes are pickled (to remove oxide remnants) and then they are tin plated in a galvanic process. In between the products are transported, rinsed and dried. There are two production lines with a common tin bath and separate pickling baths.

The PAT 'SNPB' derived its name from the assignment ('Snel Naar Proces Beheersing', in English: 'quickly to process control), and from two important chemicals of the tin bath: tin (Sn) and lead (Pb). To monitor the process the operators use more than ten measurements. The large tin bath is chemically analysed twice a week, both pickling baths every shift. A computer is fed these results and gives replenishment advises for the baths. At the same time the thickness of the tin layer is monitored by measuring a sample of the products of each shift.

The PAT started improving the conditions in their production area. Thus a good climate was created for process improvements, which came readily. The evaporation of the tin bath was greatly reduced by covering the surface, the homogeneity of the bath was improved by installation of mixers, devices to replenish the baths at the bottom reduced air inclusion, *etc.*. The PAT also managed to reduce the waste of chemicals by recycling used chemicals and devising a new replenishment scheme. To secure the improvements the main operators of this process were accompanied in the PAT with their substitutes when the control charts and new procedures were implemented. The work instructions were updated by a few operators who rewrote them in their own language. The arrangements with mechanics about preventive maintenance were also improved. The final result is that the operators have pride again in their work and their work place, and they want to keep it that way! Additionally, the process owner saves $50,000 yearly, owing to less waste of chemicals—apart from the process being now under control. The process capability study of this process showed this result. Here we only discuss the concentration of tin in the tin bath, and the layer thickness on the diode.

Tin Concentration (in the Tin Bath)

The process instruction requires a tin concentration between 100 and 160 g/l. Figure 5.8 displays the control chart of concentrations of 33 analyses of the tin bath of the last three months. The control limits (LCL = 111.7 g/l, UCL = 140.4 g/l) were calculated with the moving range method of step 7. The chart shows that the tin concentration is under control. There are not enough measurements to calculate a process capability index, but we can establish that the control limits are well within the specification limits (a meaningful remark since the control chart deals with individual measurements, not with averages).

Layer Thickness (on the Diode)

The layer thicknesses of products, assembled around the same three months period, are investigated from 177 samples. Each sample consists of 7 products, and the layer thickness of both leads of a diode was measured. The control chart of the mean layer thickness is displayed in Figure 5.9. Four averages (*i.e.*, 2.6%) exceed the control limits (LCL = 1.20 μm; UCL = 3.13 μm). The control chart of the standard deviation of the thickness (Figure 5.10) shows that all points are within the control limits (LCL = 0.05 μm; UCL = 1.41 μm). So we conclude that the layer thickness is well under control.

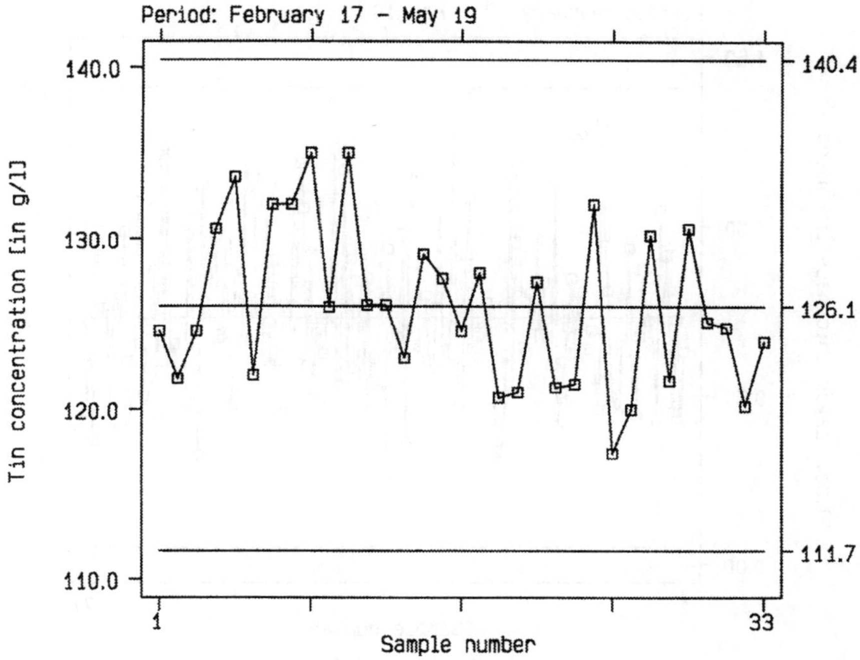

Figure 5.8: Control chart of tin concentration.

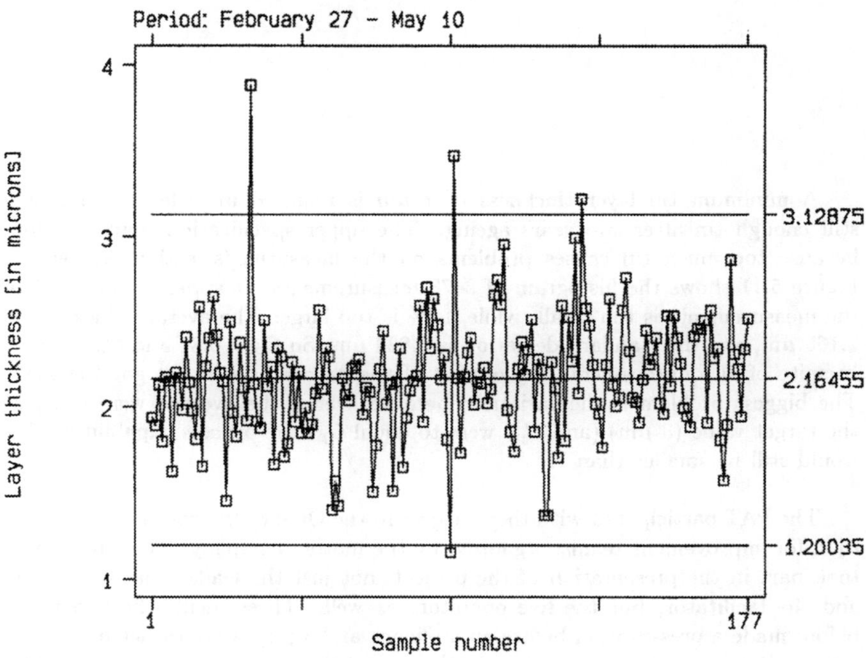

Figure 5.9: Control chart of the mean layer thickness.

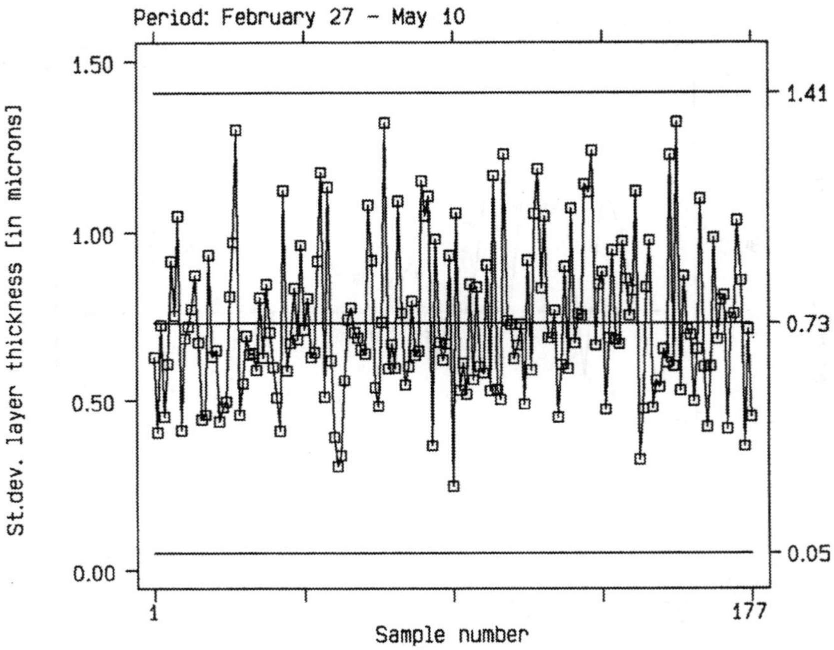

Figure 5.10: Control chart of standard deviation of layer thickness.

A minimum tin layer thickness of 1 μm is required in order to guarantee still enough tin after two years ageing. The upper specification limit is 5 μm, because too much tin causes problems on the measuring/selection equipment. Figure 5.11 shows the histogram of 2478 measurements. It appears that 2.3% of the measurements is too small, while 0.4% is too large. The average thickness is 2.165 μm, and the standard deviation is 0.824 μm. So $C_p = 0.81$ and $C_{pk} = 0.47$. In spite of all improvements the quality of the process is still not good enough. The biggest problem is the variation, because even if the average were to equal the target value (3 μm) (and C_{pk} were to equal C_p) the process capability index would still be smaller than 1.

The PAT participated with their project in the Quality Improvement Competition for improvement teams, organized by the mother company. The whole team took part in the presentation of the project, not just the leader, the supervisor, and the facilitator, but the five operators as well. These men, who had never before made a presentation before an audience and a jury, were rather nervous of course. But it was very instructive to them, and they were proud on their public presentation, as well as of their achievements in the team. Even though the jury, chaired by the plant manager, gave the prize to another team.

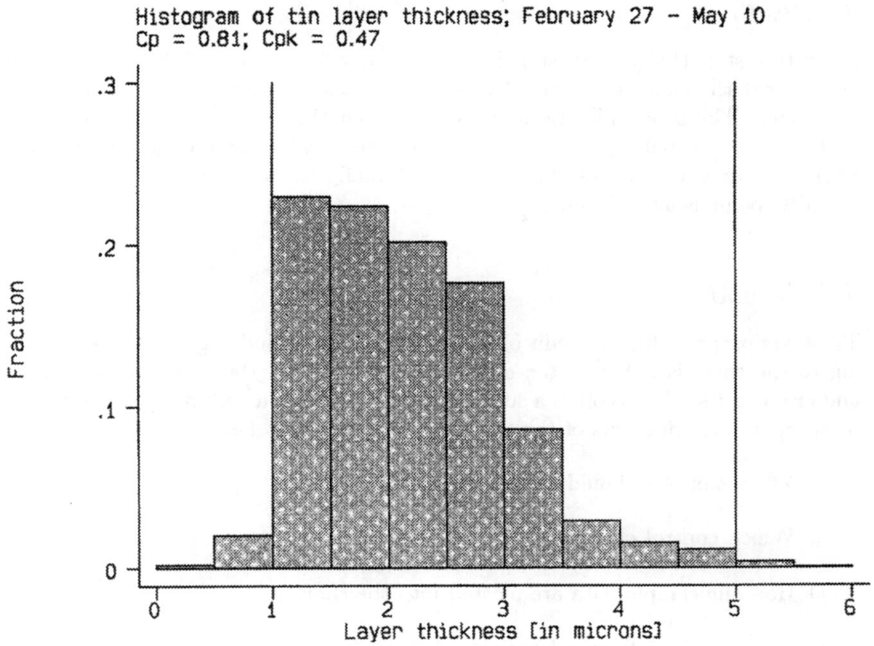

Figure 5.11: The measured layer thicknesses compared with the specifications.

5·11 Certification (Step 10)

1 Summary

The work of the team and the performance of the SPC point are judged by the commissioner. The checklist of Frame 4.4 could be used for an audit. In this final step of the activity plan the team prepares for this audit, and devises a scheme for the maintenance and further improvement of the SPC point.

2 Goal

The work of the team is judged according to fixed criteria, so that the team knows what is expected and the commissioner knows what end result he should expect. The certification is also a suitable moment to formally conclude the assignment and to thank the PAT members for their work. But the work on the process step is not finished! Continuous improvement implies continuous activities. It should be ensured, therefore, that the SPC point will be monitored, and that control limits and OCAP are updated regularly. It should be avoided that some day it will be discovered that one might as well start all over again.

3 *Result*

After this step the process step is a certified, acknowledged SPC point. This means that the elements of an SPC point are installed and implemented on the shop floor. The process is also under control, but this does not necessarily mean that it is also capable. The results are documented in operating instructions, FMEA documents, control plans, *etc.*. And finally the structure for assurance of the SPC point is agreed upon.

4 *Method*

The team prepares for the audit of the commissioner by judging its results according to the checklist. Before the commissioner is invited the team can dot its i's and cross its t's. This is often a suitable moment to give a dedicated SPC training to all operators. Subjects of the training program might be:

- ☐ When samples should be taken;
- ☐ Which control charts have to be used;
- ☐ How the sample data are plotted into the charts;
- ☐ The meaning of control limits;
- ☐ The difference between control limits and specification limits;
- ☐ The use of the out of control action plan;
- ☐ How, where, and why to record corrective actions.

In section 4.7 we discussed the audit in more detail. And in Chapter 10 the assurance of an SPC point is dealt with in more detail.

5 *Case Illustration*

The PAT 'Measure it well, know it better' has already been introduced in the case illustration of step 1. After about a year the team was certified and its assignment accomplished. During that year the team worked according to the activity plan and implemented several important improvements in the measuring/selection process. Here we mention just a few:

- ☐ The concept of the measuring contacts is improved;
- ☐ Labels on the boxes (yield and rejects) are much better readable;
- ☐ The communication with the maintenance mechanics is much better;
- ☐ A small handbook has been made with the most important instructions to operate the equipment.

Several procedures have been sharpened, and the chances of errors have been diminished. All improvements were communicated carefully to all operators. To this aim the team made a booklet with all activities and the new arrangements and procedures, and during meetings with all operators the booklet was elucidated by PAT members.

To monitor the selection process the team designed a system with control charts (this system differs from the system with 'gauge diodes', as designed by the team in the case illustration of step 4). All types of diodes have their own typical yield (based on data of the last six months), with lower and upper limits. For some types of reject categories also lower and upper limits were determined. The following procedure is now adopted: from every batch (over 40000 products) the yield is calculated after some 1000 products have been measured/selected. This value is plotted in the control chart, and when the limits are exceeded the OCAP is used to check whether the equipment is to blame, or whether the diodes come from a rather bad (or good) batch of crystals.

The operators checked the procedure extensively, and the OCAP was revised a few times. Only then the commissioner (the factory manager) was invited to check the results of the team. He in his turn asked the quality manager to do a formal audit on the basis of the checklist of Chapter 4. The results were communicated in a report, and were elucidated in a meeting with all operators and the commissioner present. In this meeting the PAT was congratulated, and awarded the certificate. All operators were treated to pastry and were given a reward for the achievements of the whole group.

6 *The Audit Report of the Quality Manager*

1 Presence of SPC elements

1.1 Is the process step monitored through control charts?
A: Yes

1.2 Are the control limits correctly evaluated?
A: Yes, historical data has been used to calculate approximately 95% limits

1.3 Are there Out of Control Action Plans?
A: Yes

1.4 Are out of control situations logged?
A: Yes, on control charts (not in a log book)

1.5 Are the specification limits well defined?
A: Not applicable

2 Use of SPC elements

2.1 Do all operators fill in the control charts correctly?
A: Yes

2.2 Do all operators recognize out of control situations?
A: Yes, only data beyond the control limits; no further rules

2.3 Do all operators know the difference between a control limit and a specification limit?
A: Not applicable

2.4 Do all operators use the OCAP in an out of control situation?
A: Yes

2.5 Are out of control situations correctly logged by all operators?
A: Yes, the code of the corresponding action is recorded on the control chart

2.6 Is process information passed on at shift changes?
A: Yes

2.7 Does the supervisor check whether control charts are filled out?
A: Yes, completeness is checked when the data are put in the computer for further analysis

3 Capability of the process step

A: Not applicable; there are no specification limits on the selection process

4 Documents control

4.1 Are operating instructions conform to the arrangements made about control charts and OCAPs?
A: Yes

4.2 Is the FMEA of the organization updated?
A: Not applicable

5 Maintenance of the SPC system

5.1 Is there a team responsible for maintenance of the SPC system?
A: The PAT continues as a Process Improvement Team

5.2 Are control limits reviewed regularly?
A: Yes, twice per year

5.3 Are the OCAPs evaluated and improved regularly?
A: Yes, it's on the agenda of the meetings with all operators (twice per year)

5.4 Is the process capability regularly evaluated?
A: Not applicable

5·12 Guidelines for Use of the Activity Plan

The activity plan of this chapter is widely applicable. In companies such as Philips Semiconductors in Stadskanaal and Hollandse Signaalapparaten in Hengelo it proved to be successful in implementing SPC. And the reader may recall that these companies are really different: mass production in Stadskanaal and small lot production Hengelo.

Implementation of SPC at ASM Lithography in Veldhoven met with difficulties, however, when the activity plan was used. The reason is that this company has extremely extended process steps with only low volume production. Furthermore, ASML is focused on research and development, and this leads to oft modified designs. Owing to the introduction of new technology the process steps are therefore often subject to change, as early during the diagnostic phase. The program of the activity plan, beginning with distinguishing small process steps, does not fit in with the company's structure. ASML makes a clear distinction between two areas. The first area consists of all assembling operations, in which technicians assemble the separate modules; these operations are considered as one big, connected process. The second area consists of integrating all separate modules into a complete wafer stepper. Both areas are so large that steps 1–3 amount to hundreds of cause and effect combinations. The throughput time of these steps can become more than six months, long enough for several process modifications. That is why ASML adapted the standard method to the needs of their own organization. This, however, is valid for any organization: adapt the standard to fit your own needs and wishes, and make it part of your own culture. The adaptation of ASML is now described:

Step A

Start with the process description (step 1 of the activity plan). Compare written procedures with the actual way of working, and bring the two into agreement (either modify the procedures, or standardize the way of working).

Step B1

Introduce a log book and log all process disturbances. Describe these disturbances through cause and effect combinations, and calculate the risk by multiplying the severity of the effect with the frequency of the cause (which should be known after three months logging). Severity and frequency are numbers from 1 to 10, as in step 3 of the standard method.

Step C1

Improve high risk cause and effect combinations (as in step 4).

Step B2

Parallel to step B1, relevant parameters (as indicated by Development) are recorded for 4–5 weeks. Tentative control limits are then calculated, and signals and actions are logged.

Step C2

After approximately 3 months of practice the tentative control charts are adapted and implemented on the shop floor, together with an OCAP.

Step D

The logging of step B1 continues meanwhile. The information of the most recent three months (a 'moving time window') is reviewed every two weeks—when the team meets and risk numbers are determined (step B1). Of course, actions are started (step C1) to lower high risks, and steps B2 and C2 are repeated for all new measurements.

Next Steps

The steps 6–10 of the activity plan (measurement analysis, control charts, OCAP. process capability study, and certification) are done according to the description in this Chapter.

The advantage of this adapted method is that the process diagnosis can still be done in 3–4 months. This ensures progress; and equally important is that process modifications are incorporated immediately.

Chapter 6

Principles of the Shewhart Type of Control Charts

6·1 Introduction

The first letter of 'SPC' stands for Statistical, referring to statistics as an essential discipline in controlling and improving processes. It has become clear from the previous chapters that effective implementation of SPC requires far more than application of selected statistical techniques. Success is reached to the extent that statistical thinking becomes natural in management of processes within a company. Adequate understanding of the statistical principles underlying the techniques commonly applied is of prime importance to reach this success. This goes well beyond calculating control limits in several standard situations or determining the C_{pk} of a process.

The control chart, specifically the Shewhart type of control chart, is probably the most widely known statistical technique associated with SPC. In Chapter 5, a basic method of establishing the appropriate limits in control charts was presented. In this chapter the general concept of the Shewhart type of control chart will be outlined. Whether or not the statistical properties of such charts should be evaluated in detail has been subject to considerable debate since they were introduced (see Neave (1992), for example). Control charts are currently applied on a large scale, and in numerous cases they determine when to stop production processes and take corrective action. Thus their actual performance in terms of false alarms and speed of detecting out of control conditions has a large economic impact. Hence the statistical properties need evaluation. This evaluation has been undertaken since the invention of the control chart, but mostly based on the assumption that distributions were normal and the parameters were known. This has been quite reasonable up to a decade or so ago, owing to the high volume production in which the methods were applied and the control chart for a subgroup mean being the favourite.

Shewhart type of control charts are introduced in section 6.2. Hillier (1964, 1967 and 1969) was one of the first to investigate the impact of much smaller volumes of available data on control chart performance. A formal framework for Shewhart type of control charts is given in section 6.3, based on his original ideas. Thereafter basic performance issues are addressed, with emphasis on performance when distributional parameters need to be estimated (6.4).

6·2 General Setting for Shewhart Type of Control Charts

A Shewhart type of control chart consists of a graph with time on the horizontal axis and a control characteristic (an individual measurement or a statistic such as the mean or range) on the vertical axis. Control limits provide the first easy checks on the stability of the process, *i.e.*, no special causes present. Three control limits are drawn: the Central Line (CL), the Lower Control Limit (LCL) and the Upper Control Limit (UCL). The limits on the control charts provide an *operational definition* for the distinction between variation owed to *common causes* and variation owed to *special causes*. A signal on the chart—such as a point above the UCL—indicates a special cause. If no signals occur the process is assumed to be under control, *i.e.*, only *common causes* of variation are present. As described in Does and Schriever (1992) control charts can be used for *past data* (examining a historical set of data) and *current control*. The focus in this book is on construction and evaluation of control charts as part of a feedback control loop in the process, hence on *current control*. When the control chart signals that a special cause was present, the Out of Control Action Plan should be activated. Of course, points outside the control limits will always present such a signal, but it is often sensible to specify additional rules. One commonly known pattern attributed to a special cause is six points in a row steadily increasing or steadily decreasing (Figure 6.1).

Effective application of control charts requires:

- □ An adequate understanding of the principles underlying control charts;
- □ Application of the charts to those characteristics and parameters of the process that contribute to control and improvement;
- □ An effective sampling scheme for the charts.

The distinction between special and common causes of variation is the central principle underlying SPC and control charts. At first sight this principle appears straightforward. In practice the distinction is seriously complicated because of a lack of knowledge of the process inherent variation (common causes). Managers invariably ask for *the* cause of the delay in the recent software project, completely ignoring that many common causes might have been at work. Similarly, in presentations of the financial results of a company the performance of the most recent period is compared with one or two previous periods. Any increase or decrease is

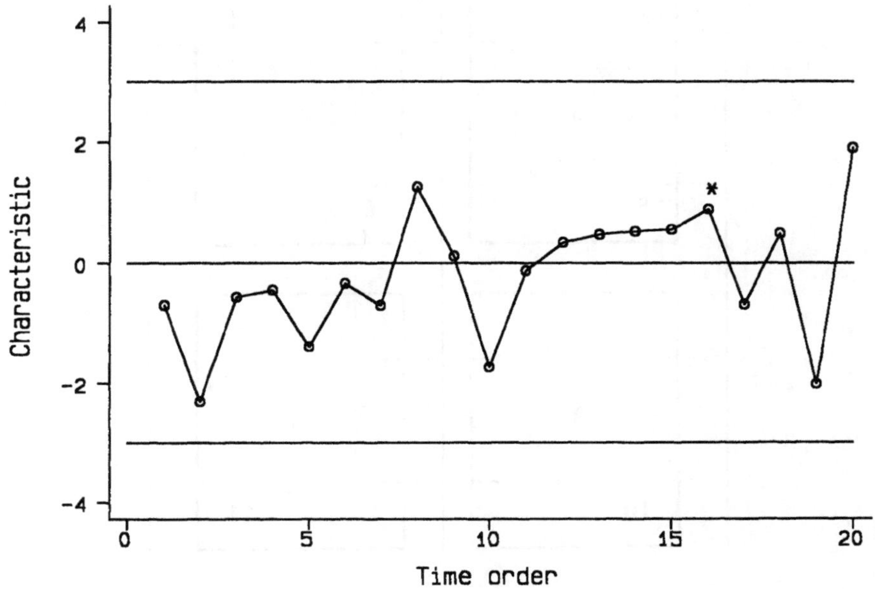

Figure 6.1: Example of a Shewhart type of control chart with signal based on 6 points in a row steadily increasing (*).

usually 'adequately' explained. Only rarely will a company admit that fluctuation of performance was within the bounds of the common cause variation and therefore cannot be explained seriously by a few causes. Ignorance of the magnitude of common cause variation at this high level of abstraction from actual business processes probably makes such assessments almost impossible.

Common cause variation may also be hard to quantify in industrial processes. An example from one of the many companies which implemented SPC along the lines of this book illustrates this.

Example

In the production process of Printed Circuit Boards (PCBs) a thin copper layer is deposited during an electroplating process step. Uniform thickness throughout the board is important for both subsequent process steps and the final product. In the electroplating process several PCBs are treated in one run. To this end, PCBs are positioned in a galvanic bath according to a fixed arrangement (Figure 6.2). The copper's thickness is mainly determined by the current density, total (metal) surface of the PCBs, process time, copper concentration and the surface of the anode.

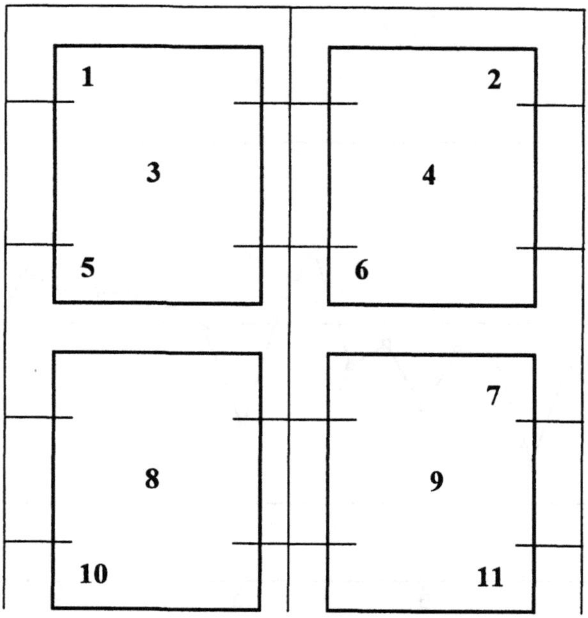

Figure 6.2: PCBs as positioned in the copper electroplating process.

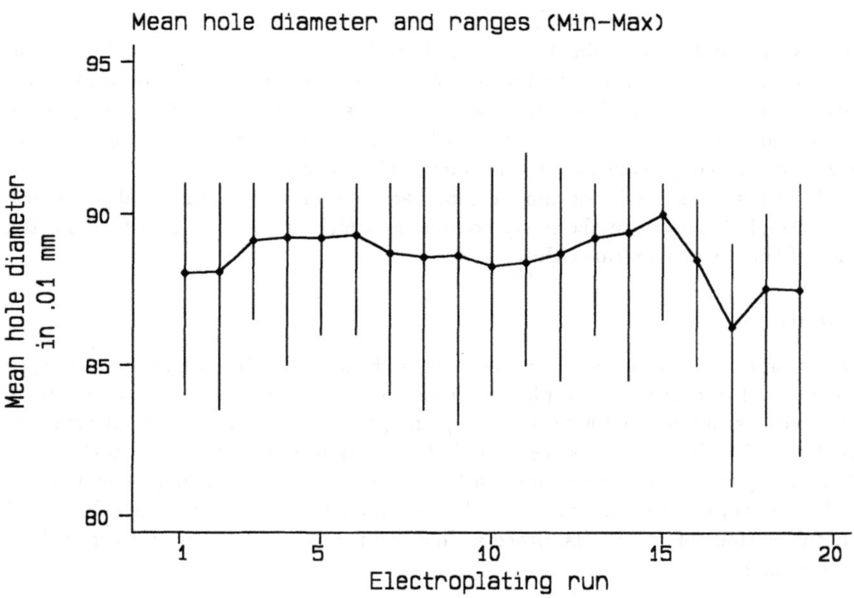

Figure 6.3: Trend chart of mean hole diameter and ranges within a run.

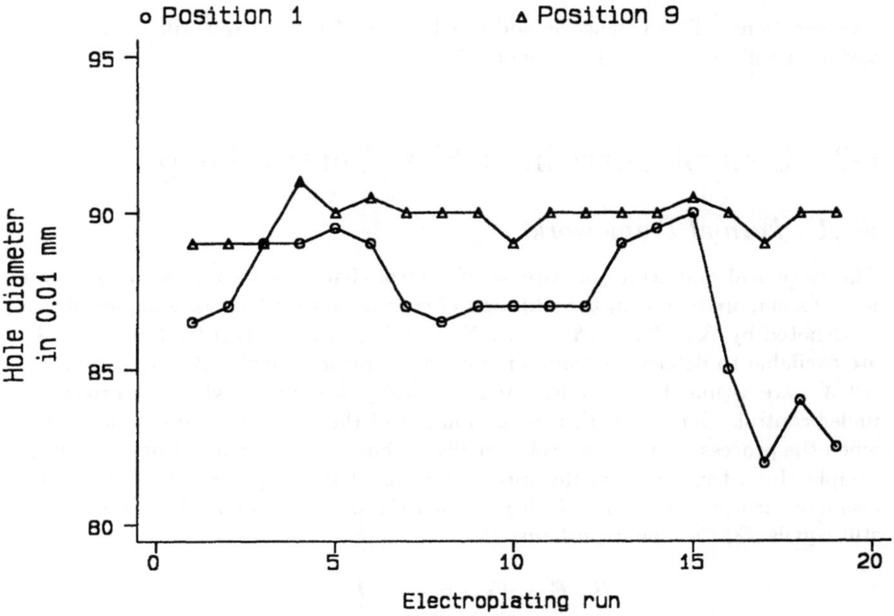

Figure 6.4: Trend chart for hole diameters at positions 1 and 9.

In this particular process, the copper's thickness can be adequately measured by determining the holes' diameters after electroplating. On various grounds the Process Action Team (PAT) decided to measure eleven holes' diameters at the indicated positions for nineteen runs in a period of three weeks.

The diameter of a bare hole is 1 mm.; on average the thickness of the copper layer is about 0.13 mm. The average diameter per run appears to be under control, with the largest variation present within a run (Figure 6.3). The process seems to be under control, but the variation within a run may need to be reduced to conform to tight requirements on these high quality products.

Closer examination of the diameters per position within a run presents some surprises: at position 1 (upper left corner) a clear jump occurred at the 15th run, which is not present at position 9 (Figure 6.4). Hence the process cannot be considered under control. However, the special cause signalled yields an opportunity for improvement: if removed, this will lead to a smaller within run variation. ■

The variation in the process needs to be studied quite thoroughly in order to reveal and quantify variation owed to special and common causes, as this example illustrates. This will also form the basis of adequate control charts. Control chart performance then depends on:

- □ How quickly it signals a special cause when one has actually occurred;
- □ How often it triggers a search for a special cause if the process is in fact under control (false alarms).

Of course, one wants the first to happen as quickly as possible and the second never. As this is impossible it requires a balance based on statistical and economic considerations. This balance is addressed in this Chapter; quantification of the variation will be addressed in Chapter 7.

6·3 Control Chart in an SPC Control Loop

6·3.1 Formal Framework

The statistical and economic aspects of control charts need to be addressed in a more formal approach. In the situation of current control k initial samples of size n (denoted by $\boldsymbol{X}_1$, $\boldsymbol{X}_2$,...,$\boldsymbol{X}_k$, with $\boldsymbol{X}_t = (X_{t1}, X_{t2}, \ldots, X_{tn})$ for $t = 1, 2, ..., k$) are available to determine limits on a chart for future samples $\boldsymbol{X}_{k+1}$, $\boldsymbol{X}_{k+2}, \ldots$. All X_{tj} are assumed independent and identically distributed when the process is under control. Denote by F_0 the distribution of the observed characteristics X_{tj} when the process is under control. Usually F_0 has to be estimated from the initial sample. In order to control the current process at time t (for $t > k$) it has to be evaluated if the distribution F_t from which the observations in $\boldsymbol{X}_t$ are sampled still equals F_0, *i.e.*, the hypothesis:

$$H : F_0 = F_{k+1} = \cdots = F_t \tag{6.1}$$

has to be tested. Commonly the distribution of F_t is known up to a few parameters (*e.g.*, location and spread) and thus for each (relevant) parameter a chart can be constructed to test hypothesis (6.1) (see Does and Schriever (1992)). When the customary assumption is made that the observations X_{tj} come (under H) from a normal distribution $N(\mu, \sigma^2)$, both μ and σ are parameters to control. For future observations parameters may have shifted for a sample number t, which indicates an out of control situation.

In general, a parameter η is controlled as follows. Let $Y_t = Y(\boldsymbol{X}_t)$ be an estimate of the parameter η_t in sample t. Furthermore, let T and V be summary statistics such that $T = T(\boldsymbol{X}_1, \boldsymbol{X}_2, \ldots, \boldsymbol{X}_k)$ is a consistent (as $k \to \infty$) estimator of η and $V = V(\boldsymbol{X}_1, \boldsymbol{X}_2, \ldots, \boldsymbol{X}_k)$ a consistent (as $k \to \infty$) estimator of the spread of the distribution of Y_t under H.

A control chart for η based on T and V is a plot of Y_t against t. In the control chart, limits of the form:

$$L = T + c(n, k, p)V \tag{6.2}$$

are drawn, where the constant $c(n, k, p)$ satisfies the probability statement:

$$\begin{aligned} P_H\{Y_t \leqslant L\} &= P_H\left\{\frac{(Y_t - T)}{V} \leqslant c(n,k,p)\right\} \\ &= p \quad (t > k). \end{aligned} \tag{6.3}$$

Hence, through (6.3) the error rate of wrongly signalling out of control situations for $\boldsymbol{X}_t$ is controlled. In many cases the distribution of $(Y_t - T)/V$ is symmetric around zero, and $c(n, k, p) = -c(n, k, 1 - p)$. The distribution of Y_t is tacitly assumed to be normal. The CL on the control chart corresponds to $p = 0.50$.

Traditional 3σ ***limits*** for the known parameters case correspond to p=0.99865 (UCL), and $p = 0.00135$ (LCL). The methods incorporated in the majority of textbooks let $c(n,k,0.9865) = 3$ irrespective of n or k. Thus the actual probability of incorrectly signalling out of control is then not controlled, in contrast to applying (6.3). For (6.3) to be used, the distribution of $(Y_t - T)/V$ has to be known or approximated, which is possible in many cases (Does and Schriever (1992) and Roes, Does and Schurink (1993)). Hillier (1964, 1967 and 1969) used similar methods for control charts for the mean and range, but this was not generally used. Usually parameters are assumed known, despite that they have to be estimated. Consequently out of control signal rates generally stated are at best approximate and substantial deviations can occur when sample sizes are moderate (see Roes, Does and Schurink (1993) or Quesenberry (1993)). In all the considerations below, the distributions of the X_{tj} and Y_t are assumed to be normal, and hence follow a frequency distribution as modelled in Figure 6.5.

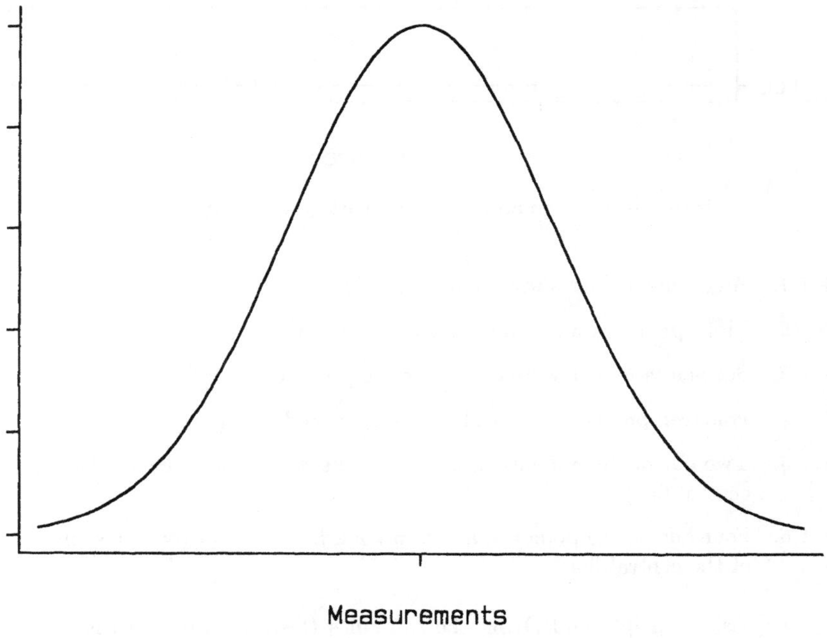

Figure 6.5: Frequency distribution of the normal distribution.

Specific run rules can be used in addition to the control limits to enhance the power of detecting out of control conditions. To this end the spaces in the chart between the central line and the upper control limit on the one side and between the central and the lower control limit on the other can both be further divided in three equal zones (Figure 6.6). On both sides of the central line these are indicated by A (closest to upper or lower control limit), B and C (closest to the central line). The following decision rules for control charts form a basic set (see Nelson (1984 and 1985) and Walker, Philpot and Clement (1991)):

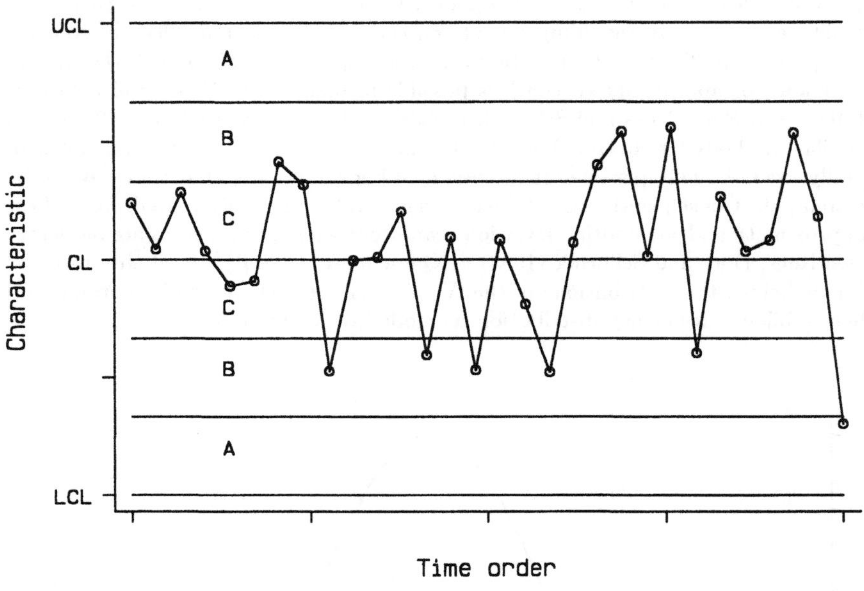

Figure 6.6: Decision areas in a Shewhart type of control chart.

Test 1. Any point falling above UCL or below LCL.

Test 2. Eight points in a row on the same side of CL.

Test 3. Six points in a row steadily increasing or decreasing.

Test 4. Fourteen points in a row alternating up and down.

Test 5. Two out of three points in a row in zone A (on the same side of the central line).

Test 6. Four out of five points in a row in zone B or beyond (on the same side of the central line).

The basic principle underlying the run rules (tests 2 through 6) is twofold. Firstly, specific patterns of out of control conditions might be detected earlier, such as a small but persistent trend. Secondly, the decision rules are designed to have roughly the same (marginal) false alarm probability under the assumption of a normal distribution with known parameters.

6·3.2 *Two Traditional Examples*

To introduce the performance evaluations of the next section, two traditional control charts are illustrated. These are the classical $\overline{X}$-chart and the chart for individual measurements with the so called 3σ ***limits***.

The Classical $\overline{X}$-Chart

Suppose k initial samples of size n are available for construction of the control chart (see section 6.3.1) and that the X_{tj} are distributed as $N(\mu, \sigma^2)$, with μ and σ unknown. Then the sample means (for $t > k$):

$$\overline{X}_t = \frac{1}{n}\sum_{j=1}^{n} X_{tj}$$

are the natural statistics to plot on a control chart for current control of μ. The central line and upper and lower control limits are customarily set using

$$\begin{aligned} \text{UCL} &= \overline{X} + 3\frac{\overline{S}}{c_4\sqrt{n}}, \\ \text{CL} &= \overline{X}, \\ \text{LCL} &= \overline{X} - 3\frac{\overline{S}}{c_4\sqrt{n}}, \end{aligned} \tag{6.4}$$

where

$$\overline{X} = \frac{1}{k}\sum_{t=1}^{k} \overline{X}_t \quad \text{and} \quad \overline{S} = \frac{1}{k}\sum_{t=1}^{k} S_t,$$

with S_t the tth sample standard deviation and c_4 such that[1] $E(S_t) = c_4\sigma$. In the formal framework of section 6.3.1, $Y_t = \overline{X}_t$, $T = \overline{X}$, $V = \overline{S}/(c_4\sqrt{n})$ and $c(n,k,p) = 3$. Note that in this case (6.3) is not fulfilled.

The Classical Chart for Individual Measurements

The case of individual measurements ($n = 1$) is special, since variability cannot be estimated on a within sample basis. Suppose, therefore, that a sample of k independent measurements $X_1, X_2, \ldots, X_k$ distributed as $N(\mu, \sigma^2)$ is available. The control chart for individuals is then constructed as follows. Let

$$\overline{X}_k = \frac{1}{k}\sum_{t=1}^{k} X_t \quad \text{and} \quad MR = \frac{1}{k-1}\sum_{t=2}^{k} |X_t - X_{t-1}|.$$

Then limits are given by:

$$\begin{aligned} \text{UCL} &= \overline{X}_k + 3\frac{MR}{d_2}, \\ \text{CL} &= \overline{X}_k \\ \text{LCL} &= \overline{X}_k - 3\frac{MR}{d_2}, \end{aligned} \tag{6.5}$$

with d_2 a constant such that $E(MR/d_2) = \sigma$. In the formal framework of section 6.3.1 $Y_t = X_t$, $T = \overline{X}_k$, $V = MR/d_2$ and $c(1,k,p) = 3$. Also in this case (6.3) is not fulfilled.

---

[1] $E(X)$ is the expected value $E(X) = \int x f(x)\,dx$, var(X) is the variance var$(X) = \int (x - E(x))^2 f(x)\,dx$, for the random variable X with probability density $f(x)$.

6·4 Performance Issues

Originally, Shewhart's choice of 3σ limits (Shewhart, 1931) did not stem from any specific mathematical computation. It seemed to him "an acceptable economic value". He did relate it to predictability within the control limits for a process that is under control. Prediction within limits meaning "that we can state, at least approximately, the probability that the observed phenomenon will fall within the given limits". Thus he motivated his choice for applying 3σ limits using two very basic results from probability theory:

- □ For any stochastic variable X with finite mean μ and standard deviation σ, there holds $P(|X - \mu| < \mathbf{a}\sigma) > 1 - 1/\mathbf{a}^2$ (Chebychev's inequality);
- □ If the probability distribution of X is strongly unimodal, this can be sharpened to: $P(|X - \mu| < \mathbf{a}\sigma) > 1 - 1/(2.25\mathbf{a}^2)$ (Camp (1922), Meidell(1922), Shewhart (1931)).

As a consequence the probability of a signal beyond these 3σ limits is below 11% for practically all probability distributions and below 5% for most distributions relevant for practical applications if the process is under control. This close link with probability and statistics has inspired much research into the performance of control charts from a statistical perspective, although some contend that such considerations are pedantic (Wheterill and Brown, 1991) since control limits depend largely on questionable assumptions. Control charts are applied at enormous scale nowadays, with large positive impact on process and product quality. On the other hand, this scale requires careful performance evaluation. Both false alarms and insensitivity to out of control conditions can lead to considerable losses, and the magnitude of these losses is directly related to the performance of charts from a statistical perspective.

The performance of control charts for current control can basically be evaluated in two (related) ways. The first is to evaluate the probability of one sample signalling out of control through a point outside the control limits. The second is to consider the run length distribution. The run length (RL) is the number of samples from the start of charting or from the start of an out of control condition until this is signalled by a point outside the control limits. In this section performance issues will be dealt with under the assumption of normal distributions for the X_{tj}.

6·4.1 Known Parameters

In case the distributional parameters (μ and σ) are known, performance can be assessed straightforwardly for the two examples above. Without loss of generality, the under control parameters μ and σ can be assumed to be 0 and 1, respectively. Out of control situations are defined as lasting jumps to a mean level of $\delta>0$ (*i.e.*, to $\mu+\delta\sigma$ in the original scale). Since parameters do not need to be estimated, both above situations can be dealt with by considering samples of size n and including the individual measurements case as sample size n=1. Then the probability of one sample showing a false alarm, P_0, is given by:

$$P_0 = 2(1 - \Phi(3)).$$

If the jump is present the probability of a signal in one affected sample, P_δ, is given by:

$$P_\delta = 1 - \Phi(3 - \delta\sqrt{n}) + \Phi(-3 - \delta\sqrt{n}) \approx 1 - \Phi(3 - \delta\sqrt{n}),$$

since the occurrence of a signal below the lower control limit can be neglected for relevant jump sizes δ. Let current control charting start at the first sample after the initial k. Then the under control run length, RL_{in}, is defined by:

$$RL_{\text{in}} = \min_{t \geqslant k+1} \left\{ t - k \mid \overline{X}_t > \frac{3}{\sqrt{n}} \vee \overline{X}_t < -\frac{3}{\sqrt{n}} \right\}. \tag{6.6}$$

In the same way, the out of control run length, RL_{out}, after a jump occurring between k_0 and $k_0 + 1 (k_0 > k)$ is defined as:

$$RL_{\text{out}} = \min_{t \geqslant k_0+1} \left\{ t - k_0 \mid \overline{X}_t > \frac{3}{\sqrt{n}} \vee \overline{X}_t < -\frac{3}{\sqrt{n}} \right\}. \tag{6.7}$$

From the probabilities P_0 and P_δ and the independence of the $\overline{X}_t$ the run length distributions can be calculated, and turn out to be geometric:

$$\begin{aligned} P\{RL_{\text{in}} = r\} &= P_0(1 - P_0)^{r-1}, \\ P\{RL_{\text{out}} = r\} &= P_\delta(1 - P_\delta)^{r-1}. \end{aligned} \tag{6.8}$$

Hence the run length distribution is completely characterized by its expectation: $E(RL_{\text{in}}) = 1/P_0$ and $E(RL_{\text{out}}) = 1/P_\delta$. This expectation is usually indicated as Average Run Length (ARL). It is convenient for several calculations to differentiate between crossing the UCL and crossing the LCL. Let $P_{0,\text{U}}$, $P_{0,\text{L}}$, $P_{\delta,\text{U}}$, $P_{\delta,\text{L}}$, $\text{ARL}_{0,\text{U}}$, $\text{ARL}_{0,\text{L}}$, $\text{ARL}_{\delta,\text{U}}$ and $\text{ARL}_{\delta,\text{L}}$ denote the probabilities and average run lengths for crossing the upper and lower control limits, respectively. Then, of course, $\text{ARL}_{x,y} = 1/P_{x,y}$ for $x = 0, \delta$ and $y = \text{U}, \text{L}$. Furthermore, $P_0 = P_{0,\text{U}} + P_{0,\text{L}}$ and $P_\delta = P_{\delta,\text{U}} + P_{\delta,\text{L}}$. Hence in this simple case it is easy to see that:

$$\begin{aligned} \frac{1}{\text{ARL}_0} &= \frac{1}{\text{ARL}_{0,\text{U}}} + \frac{1}{\text{ARL}_{0,\text{L}}} \\ \frac{1}{\text{ARL}_\delta} &= \frac{1}{\text{ARL}_{\delta,\text{U}}} + \frac{1}{\text{ARL}_{\delta,\text{L}}}. \end{aligned} \tag{6.9}$$

This relation (6.10) also holds in more complicated applications, such as cumulative sum upper and lower decision schemes (Wetherill and Brown (1991)).

To enhance the power of Shewhart type of control charts, the use of additional tests as defined in the previous section should be considered. It has been shown that simultaneous use of several tests markedly increases the false alarm rates (Champ and Woodall (1987), Walker *et al.* (1991) and Does and Schriever (1992)), or equivalently reduces the average run length when the process is under control. This is illustrated in Table 6.1 for selected combinations of test.

As to the ability to detect specific out of control conditions, Davis and Woodall (1988) have shown that the trend rule (test 3) does not enhance performance under a linear trend. Hence the purpose it is designed for is not served. Experience and

Table 6.1: Run length characteristics for simultaneous use of tests in an under control process.

Simultaneous Use of tests	ARL	Prob[$RL_{\text{in}} \leqslant 20$]
Test 1	370	5.3%
Tests 1 and 2	153	10.5%
Tests 1 and 3	200	8.9%
Tests 1 and 5	225	8.5%
Tests 1 and 6	166	10.5%
Tests 1,2,3,5 and 6	79	20.0%

further research (Quesenberry, 1995) have demonstrated that especially extension with rules 4 and 5 and possibly 2 provide an adequate balance between the under control behaviour and sensitivity to signalling shifts in the process mean. In general it is recommended to use additional run rules in addition to the 'out of limits' test with reservation. It is, in fact, only sensible, if the Out of Control Action Plan can be adjusted to match the different tests used. If this is not possible, then preferably additional run rules should not be used. It remains important, however, that the control charts are inspected for unexpected patterns.

6·4.2 *Parameters Estimated*

In general, the amount of available data does not justify the assumption that distributional parameters are known. The estimation of parameters introduces two issues which affect the performance of the charts. The first is related to (6.3). If the fixed constant $c(n,k,p) = 3$ for the UCL is applied, then $P_H\{Y_t > \text{UCL}\} = P_H\{(Y_t - T)/V > 3\}$ will be different from the $p = 0.00135$ intended. Likewise, $P_H\{Y_t < \text{LCL}\} = P_H\{(Y_t - T)/V < -3\}$ will be different. Secondly, for given s and t, the events $\{(Y_s - T)/V > 3\}$ and $\{(Y_t - T)/V > 3\}$ are no longer independent, because of the fact that in both events the statistics T and V are involved. Exact evaluation of the run length distribution thus becomes intractable, even in the case that the observations follow a normal distribution. In Quesenberry (1993) the effects of the available sample sizes on control chart performance for the sample mean and individual measurements were investigated using simulation. In Figure 6.7 the under control cumulative run length distribution of the chart for sample means is shown, based on samples of five. The geometric distribution for the known parameters case is shown, as well as the (simulated) distribution based on estimated parameters from twenty samples of five.

The difference between the two distributions is not very large. The most important feature is the fact that the two lines cross. Up to about 500 runs, the chart based on estimates is more likely to signal false alarm; above 500 runs the situation is reversed. Similar results are obtained when the chart for individual measurements is evaluated (Figure 6.8). In this case, the difference between the cumulative distributions is larger. The lines cross between runs 250 and 350, depending on the sample size. Based on these and similar results for a range of sample sizes, Quesenberry (1993) recommends a minimum of 100 samples of 5 be-

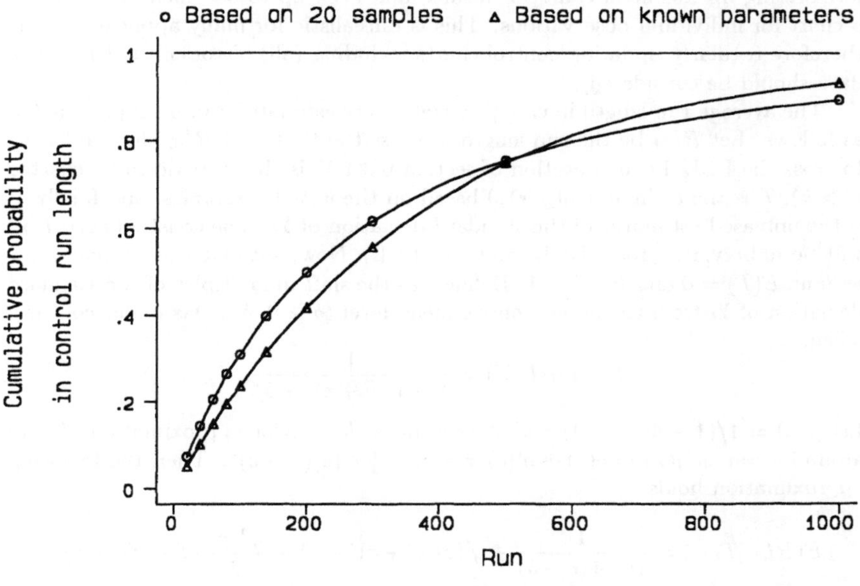

Figure 6.7: Under control run length distribution for current control on samples of five.

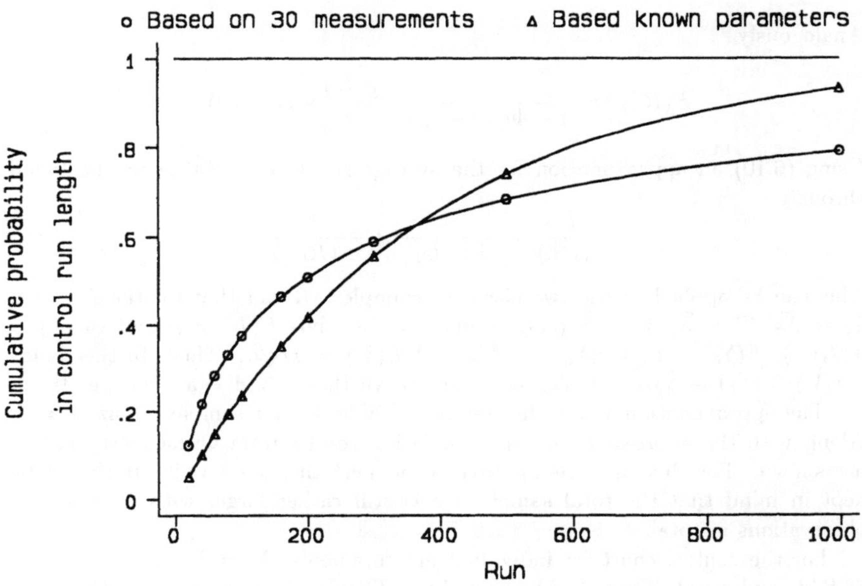

Figure 6.8: Run length distribution for current control on individual measurements.

fore setting up a control chart for means, and even up to 300 measurements for a chart for individual observations. This is unrealistic for many applications and therefore regularly updating control charts including (all) historical under control data should be considered.

The average run length in case parameters are estimated can be approximated as follows. Let RL_U be the run length to cross the UCL and RL_L the run length to cross the LCL. In the notation of section 6.3.1 Y_t is the statistic to be charted $(t > k)$, T is the estimator of $E(Y_t)$ based on the initial k samples, and finally V is the unbiased estimator of the standard deviation of Y_t. The constants $c(n, k, p)$ will be abbreviated to c (UCL) and $-c$ (LCL). Now, without loss of generality, assume $E(T) = 0$ and $E(V) = 1$. Define δ as the shift in multiples of the standard deviation of Y_t from the under control mean level ($\delta \geqslant 0$, $\delta = 0$ is under control). Then:

$$E(RL_U|T, V) = \frac{1}{1 - \Phi(T + cV - \delta)}.$$

Let $g(x) = 1/(1 - \Phi(x - \delta))$ and the second order Taylor approximation of $g(x)$ around $x = c$ is $g(x) = g(c) + g'(c)(x - c) + \frac{1}{2}g''(c)(x - c)^2$. Then the following approximation holds:

$$E(RL_U|T, V) = \frac{1}{1 - \Phi(c - \delta)} + g'(c)(T + cV - c) + \frac{g''(c)}{2}(T + cV - c)^2.$$

The unconditional expectation of RL_U can thus be approximated as:

$$E(RL_U) = \frac{1}{1 - \Phi(c - \delta)} + \frac{g''(c)}{2}\,\mathrm{var}(T + cV).$$

Analogously,

$$E(RL_L) = \frac{1}{1 - \Phi(-c - \delta)} + \frac{g''(c)}{2}\,\mathrm{var}(T - cV).$$

Using (6.10) an approximation for the average run length (ARL) can be found through:

$$\frac{1}{E(RL)} = \frac{1}{E(RL_U)} + \frac{1}{E(RL_L)}.$$

This can be applied to the two classical examples. Recall that for the $\overline{X}$ chart, $Y_t = \overline{X}_t$, $T = \overline{\overline{X}}$, $V = \overline{S}/(c_4\sqrt{n})$ and $c = 3$. Thus $E(T) = \mu$ and $\mathrm{var}(T) = \sigma^2/(nk)$, $E(Y_t) = \mu$, $\mathrm{var}(Y_t) = \sigma^2/n$ and $E(V) = \sigma/\sqrt{n}$. There further holds $\mathrm{var}(V) = \sigma^2(1 - c_4^2)/k$. Finally, it is well known that T and V are independent.

The approximation yields the results in Table 6.2, for samples of size $n = 5$. Along with the approximation, the simulation results from Quesenberry (1993) are shown. For this case the approximation performs quite well. It should be kept in mind that the total sample size is still rather large, with at least 150 observations in total.

For the control chart for individual measurements, $Y_t = X_t$, $T = \overline{X}_k$, $V = MR/d_2$ and $c = 3$. Thus, $E(T) = \mu$ and $\mathrm{var}(T) = \sigma^2/k$, $E(Y_t) = \mu$, $\mathrm{var}(Y_t) = \sigma^2$ and $E(V) = \sigma$. It further holds that $\mathrm{var}(V) \approx \sigma^2(0.8264k - 1.082)/(k-1)^2$ (Cryer and Ryan, 1990). Finally, T and V are independent in the case where the X_t are independent, identically distributed as $N(\mu, \sigma^2)$. This follows from a result in Leh-

Table 6.2: ARL with estimated control limits for the mean based on k samples of 5.

k	Method	Jump δ			
		0	1	1.5	2
30	Simulation	403	5.03	1.60	1.09
	Approximation	402	4.58	1.58	1.08
50	Simulation	389	4.73	1.63	1.09
	Approximation	389	4.55	1.57	1.08
75	Simulation	384	4.47	1.58	1.08
	Approximation	383	4.53	1.57	1.08
100	Simulation	381	4.50	1.58	1.08
	Approximation	380	4.52	1.57	1.08
200	Simulation	368	4.56	1.53	1.08
	Approximation	375	4.51	1.57	1.08
∞	Both	370	4.50	1.57	1.08

Table 6.3: ARL with estimated control limits based on k individual measurements.

k	Method	Jump δ		
		0	1	2
50	Simulation	1106	91.2	8.70
	Approximation	740	68.4	7.99
75	Simulation	663	68.3	8.15
	Approximation	617	60.2	7.42
100	Simulation	574	59.8	7.16
	Approximation	554	56.1	7.14
1000	Simulation	388	45.5	6.46
	Approximation	389	45.1	6.39
∞	Both	370	44.0	6.30

mann (1986, p. 191). The approximation results are shown in Table 6.3, along with the simulation results from Quesenberry (1993). It is clear from Table 6.3 that the approximation for the ARL performs badly for moderate, but realistic, sample sizes. From a sample size of 75 onwards the approximation is accurate enough, which is still well below the sample sizes recommended by Quesenberry (1993) as the minimum for setting up the control chart.

It is particularly clear from the evaluation of performance for known parameters that large samples are needed in order to establish reliable control limits: at least about 100 for individual measurements and at least 20 to 30 samples of size 5 for the classical case of the $\overline{X} - S$ chart.

Chapter 7

Designing Control Charts to Support Improvement

7·1 Introduction

Control and improvement of a process step critically depends on the proper choice of the characteristics plotted in control charts. The diagnostic phases covered by the Process Action Team (Chapter 5) are tuned to optimise choices for a particular process. The thickness of silicon wafers after grinding, and glass bead diameters after glass deposition in the assembly of diodes, are such characteristics. Once the characteristics have been decided upon, sample frequency and structure—*e.g.*, position(s) in machine, position(s) on products—have to be determined. The following structured approach (Roes and Does (1995), Does, Roes and Trip (1995)) can be implemented to arrive at useful sampling schemes for control charts:

1 Monitor the process intensively for an appropriate period of time.

2 Construct an appropriate model for the common cause variation, based on the nature of the process and supported by the data.

3 Design the sampling plan and establish the characteristics for control charting.

4 Design the control chart(s).

In this chapter the approach is explained in more detail, and the statistical methods applicable are introduced. Data collection, measurement scales, and (graphical) exploration of the resulting data are discussed to support the first step. Model construction based on distinguishing appropriate variance components is dealt with next. Finally, sampling schemes and control chart layout will be discussed. Actual statistical design of Shewhart type of control charts is covered in Chapter 8.

7·2 Data Collection and Measurement Scales

7·2.1 Data Collection

In the diagnostic phase of the Process Action Team, data is usually collected to gain insight into the magnitude and causes of the major components of variation of the process. Quite often data has already been gathered on a routine basis. At first glance these data might very well be used for diagnostic purposes. However, the purpose of gaining the insight mentioned is not always served best by routinely collected data. To set up useful data collection it is important to consider the basic process model in Figure 7.1.

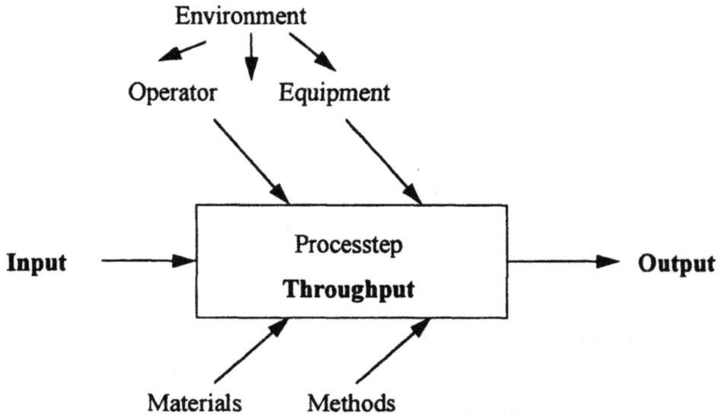

Figure 7.1: Process model.

In the process step considered *input*, *throughput*, and *output* can be distinguished. Each of these parts contributes to the total variation: incoming semi-finished products and components, process settings, operators, and many other factors affecting throughput, and measurement variation. For diagnostic purposes data needs to be collected on all three elements. To be able to draw unambiguous conclusions it is necessary that these measurements are collected coherently: measurements on the output of the process (product) need to be related to process conditions for that specific product, and the specific semi-finished product and other components and materials used. Designing data collection can be facilitated by following a product backwards through the process, and noting the important time points of relevant measurements and other data.

Example: 'Hot Air Levelling' of PCBs

One of the final phases in the manufacturing of Printed Circuit Boards (PCBs) is 'Hot Air Levelling'. In this process step an additional thin layer of tin is deposited to improve solderability. To this end a PCB is first sprayed with solder flux. Next it is very briefly dipped in a fluid tin bath. When the PCB is lifted from the bath the deposited tin layer is levelled by blowing hot air under high pressure lateral to the PCB (Hot Air Levelling, HAL). Finally, the PCB undergoes a cleaning process (Figure 7.2).

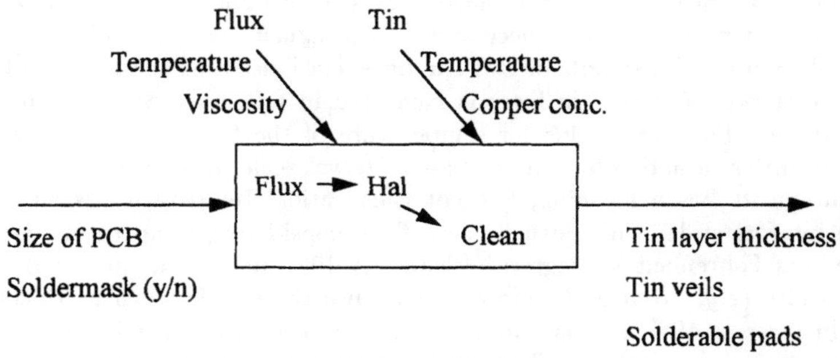

Figure 7.2: Process model of 'Hot Air Levelling'.

The FMEA has indicated the important components and characteristics contributing to variation in solderability, the essential quality characteristic after HAL. Categorised in input, throughput, and output these are:

Input Size of the PCB and the presence/absence of a solder mask.

Throughput Temperature, pH, and viscosity of the flux, temperature, and copper concentration of tin bath. Air pressure and some other process conditions are kept constant.

Output The PCBs are evaluated on a range of characteristics related to solderability and visual quality. These include: thickness of the thin layer, number of insufficiently covered pads (bad solderability), presence of tin veils and presence of stains owed to drying.

Data collection in this situation is ideally structured in such a way that for each (sampled) PCB the characteristics and parameters mentioned are registered specifically for the respective PCB. Thus, for a particular PCB its size, presence of solder mask is known, of course, as well as the quality characteristics measured after HAL. But also for that same PCB we register viscosity, pH, and temperature of the flux at the time this PCB was processed with flux, as well as temperature and copper concentration of the tin bath at the time of HAL of the PCB. In practice, this is not always easily accomplished. In this particular case, it is known that the copper concentration increases very slowly. Measurement of this concentration is difficult, and needs to be done by a specialised laboratory. Hence, it was decided to perform this measurement only once each month. ■

Data collection based on these principles usually requires a substantial effort. The pay back is substantial: unambiguous clues on the influence of many characteristics and parameters on process variability. This is crucial for the quality of the diagnostic phase and leads to well informed choices on parameters to be controlled, sampling schemes, control charts and OCAPs. Once important components of variation have been identified data collection for SPC control loop purposes will be much less demanding.

7·2.2 Measurement Scales

Measurements, inspections, and visual examination of products and processes result in data that can be statistically analysed or parameters that need to be controlled by means of control charts. To decide upon the proper analysis or control chart procedure, it is necessary to distinguish between different types of data. The previous example of Hot Air Levelling illustrates these types of data.

The thickness of the tin layer is measured in micrometers, *i.e.*, *numerically continuous*. The same holds for temperature of the flux. One can distinguish between *interval* and *ratio* scales. In an interval scale, a difference between two measurements has a meaning, but not their ratio. Temperature is an example of an interval scale, which can be seen from considering temperature measured in degrees Fahrenheit or degrees Celsius. A 10% increase is quite different on both scales (*e.g.*, from 0 °C and 32 °F). On a ratio scale, such as thickness of the tin layer, a 10% increase implies the same thickness, whether expressed in micrometers or millimetres. The crucial difference is the fact that a ratio scale has a meaningful zero, whereas the zero on an interval scale has an arbitrary position.

The presence of tin veils is categorised into one of the following four classes: none, some but not affecting quality, affecting quality and seriously affecting quality. Thus this leads to *qualitative* data, in this case *ordered categorical* data.

Finally, solder pads are sufficiently covered or not (yes/no). Such *nominal* data directly leads to counts, in this case of the number of not sufficiently covered pads ('no') per PCB and per sample of PCBs.

In general, numerical continuous data will provide the most precise and objective information. Whenever possible, such measurements should be used for control charts. Ordered categorical and nominal data are less precise and therefore require larger sample sizes to obtain the same sensitivity to out of control conditions. In this chapter we will basically distinguish between numerical continuous and nominal data. Ordered categorical data are less frequently used for control charts. Moreover, statistical theory for establishing control charts for such data is still under development. Useful charts can possibly be developed based on fuzzy set theory (Laviolette, Seaman, Barrett and Woodall (1995)), or generalised linear models (see Myers and Montgomery (1997) for an introduction). At this stage general useful recipes cannot be given.

7·3 Exploring Process Variation

7·3.1 Graphical Description of Process Variation

The data collected during the diagnostic phase of the PAT is a valuable source of information on the common cause and special cause variation of the process. This insight is necessary to establish control charts and design an adequate Out of Control Action Plan. Descriptive statistics, graphics, and more advanced analyses such as analysis of variance, regression analysis and principal components analysis will often prove useful in this respect. The more advanced topics will not be dealt with here; we refer to introductions by Montgomery (1991), Box, Hunter and Hunter (1987), Chatfield and Collins (1980) and Manly (1986). Graphical exploration of the data to distinguish variance components can prove quite powerful and is illustrated with examples in this section. Most graphical techniques are well known and will therefore not be treated in great detail.

Example: Reticle Handling Elevator

ASM Lithography in Veldhoven, the Netherlands, was referred to in Chapter 1 as a leading edge technology manufacturer of advanced wafer steppers for integrated circuit (IC) production. Wafer steppers create a pattern defining the ICs on silicon material, the so called wafers. The stepper exposes the wafer, covered with a photosensitive layer, to light or laser light projected through a reticle (a mask) which contains the pattern. The unexposed material is etched off, leaving just the circuit. Repeated application of the steps eventually defines the complicated pattern of ICs. The precision of the projection of wafer steppers is crucial for the production of circuits with ever increasing integration and decreasing size. The production volume of these complex steppers at ASM Lithography is currently about 200 per year.

In each stepper a reticle handling elevator is mounted, which in operation transports the reticle to its position. The test developed to evaluate the proper operation of the mounted elevator consists of measuring the time a defined weight needs to move the arms of the elevator from full up position to full down. These measured times are used to control the process step of mounting the elevator. In Table 7.1 the results of 40 successively mounted elevators are presented.

Table 7.1: Fall time results of 40 reticle handling elevators.

Elevator	Time (s)	Elevator	Time (s)	Elevator	Time (s)	Elevator	Time (s)
1	2.12	11	3.69	21	2.57	31	2.78
2	1.81	12	2.10	22	3.00	32	3.09
3	1.93	13	2.05	23	3.63	33	5.41
4	3.46	14	2.60	24	2.38	34	2.41
5	3.50	15	1.88	25	2.78	35	3.65
6	2.25	16	2.28	26	2.75	36	3.78
7	2.44	17	2.90	27	1.43	37	3.60
8	2.85	18	1.40	28	2.00	38	3.15
9	1.75	19	1.90	29	3.06	39	4.25
10	2.35	20	3.60	30	2.37	40	4.44

This type of data is quite uncomplicated, and can be summarised in a few basic ways: histogram, plot of individual measurements, box and whisker plot, and a trend chart. These are presented jointly in Figure 7.3.

The box and whisker plot, or box plot, is a more compact graphical summary of the data compared to the histogram. The median (F_{50}) is indicated by the central vertical line in the box and the lower and upper quartiles (F_{25} and F_{75}) correspond to the vertical ends of the box. The upwards whisker is drawn from the upper quartile F_{75} up to the largest data point just below $F_{75} + 1.5(F_{75} - F_{25})$.

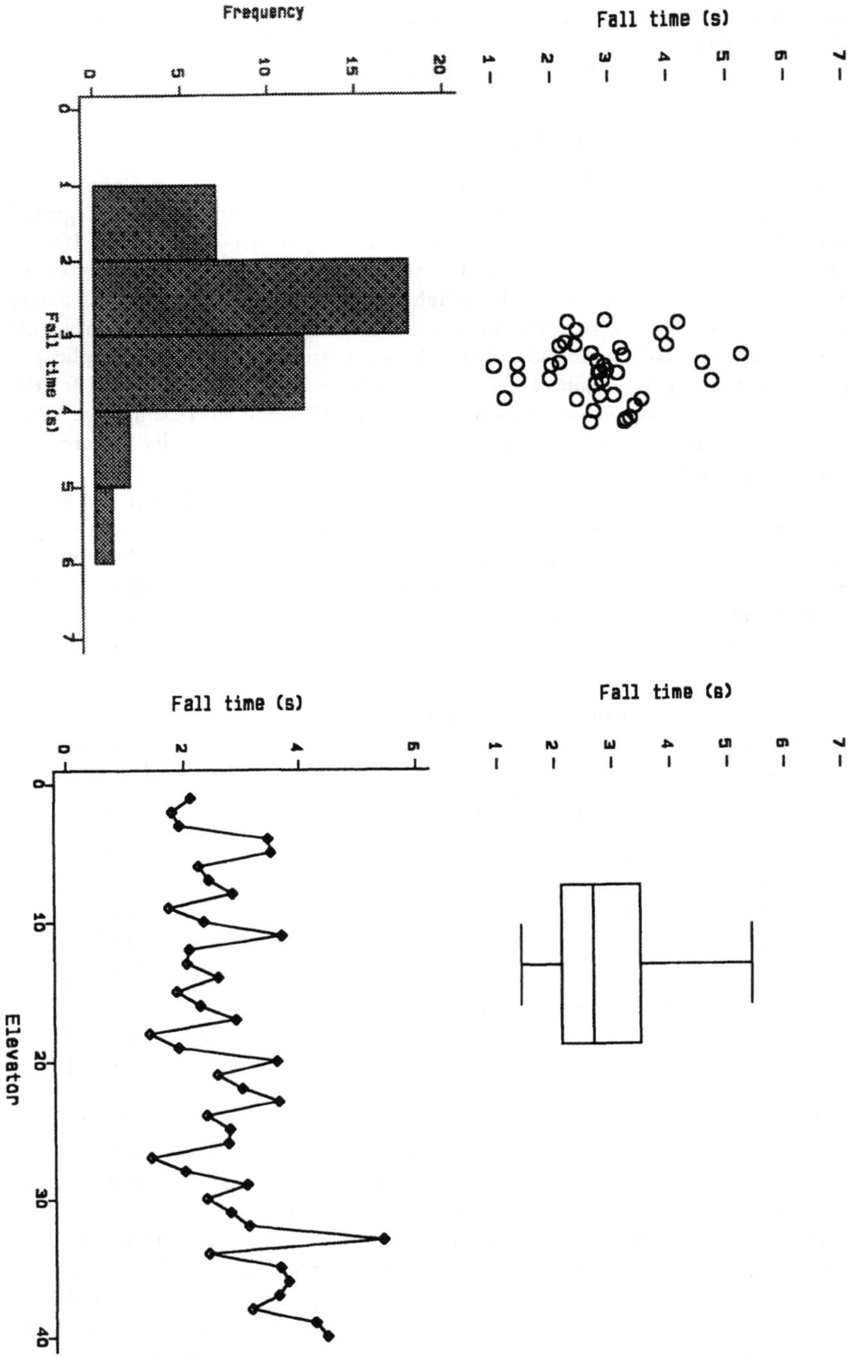

Figure 7.3: Graphical analysis of fall time data reticle handling elevators.

Similarly, the downwards whisker is drawn up to the smallest data point above $F_{25} - 1.5(F_{75} - F_{25})$. The distance $(F_{75} - F_{25})$ is called the ***inter-quartile range*** and provides a robust estimator of the spread of the observations. Data points beyond the whiskers are drawn individually and are designated outliers.

The histogram, individuals plot, and box plot demonstrate a homogeneous distribution of fall times, somewhat skewed to the right and without particular outliers indicating special causes. The trend chart (or run chart) of fall times against the order of production proves to be more revealing. The thirtieth mounted elevator stands out as an outlying observation indicating a possible special cause. Particularly the display of data against time order reveals special causes in this way, which forms the basis of the control chart.

7·3.2 Distinguishing Variance Components

The basic graphical displays of the previous section suffice for data not resulting from various sources of variation. Usually several sources of variation are at work simultaneously. In such cases, it is important to be able to identify and quantify these sources adequately. Examples include:

- Systematic differences between shifts of operators;
- Differences between positions on a mould;
- Spatial variation in chemical batches;
- Systematic drift due to tool wear;
- Random variation between and within batches.

These components of variation actually present in the process need to be identified and quantified as basis for a useful distinction between special and common causes. Control charts can then be designed to control all relevant components of the common cause system, on the basis of ***rational sampling***: sampling plans that are purposely not random and aim to describe the important variance components (Palm (1992)). The sampled subgroups for these processes will show systematic variation within subgroups and occasionally additional between-subgroup variation. The methodology to determine control limits for appropriate subgroup statistics can be more involved, but Chapter 8 provides basic tools that can be widely applied.

The graphical displays introduced in the previous section can be very well used as first method for gaining insight, if applied intelligently to subgroups defining the components of variance.

Example: Grinding Silicon Wafers

Philips Semiconductors is Europe's leading manufacturer of integrated circuits (ICs). The example presented here originates from the Philips plant in the Netherlands introduced in Chapter 1, at Stadskanaal. In Stadskanaal semiconductor crystals are manufactured on silicon wafers, which are ground to a target thickness at an early stage of the production process. The thickness of the wafer determines the height of the crystals, and therefore the electrical characteristics of the diode. Typical target values for the wafer thickness are in the range 200–300 μm. In our application the wafers are ground in batches of 31 simultaneously, positioned on so called 'frits' under the grindstone according to Figure 7.4. They are kept in place by means of vacuum decompression.

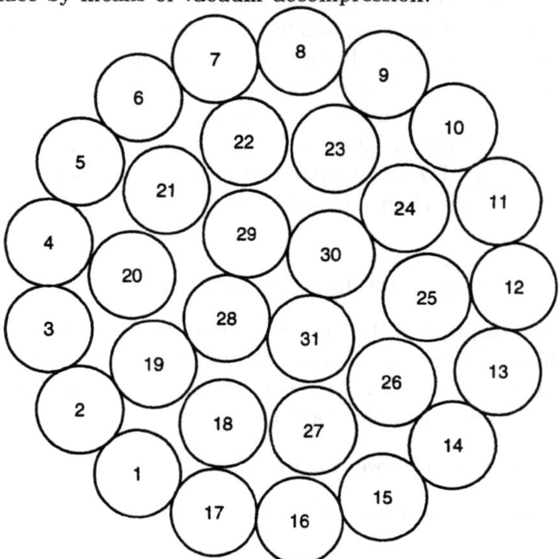

Figure 7.4: Positions of silicon wafers on the grinder.

In this example the responsible PAT arranged the process to be monitored intensively. During a period of two weeks the thicknesses of all 31 wafers of the batches processed were measured. Normally only wafers 1 and 2 from the outer circle, 18 and 19 (middle circle) and 28 (inner circle) are sampled. The grinder always stops turning in the same position, and then half of the circle with wafers is covered by a protective cap. Thus routine sampling is strongly facilitated by taking it from the uncovered half of the total grinding table. For further convenience the sample is taken from the easiest corner for the operator. These samples are used as an illustration here, with data from 30 batches displayed in Table 7.2.

In the analysis of the data from Table 7.2 we follow a stepwise approach to illustrate possible pitfalls when the specific variance structure of the process is ignored.

As a first step the histogram of all 150 measurements is drawn (Figure 7.5). The distribution of measurements seems symmetric around the mean (245 μm) and consistent with a normal distribution. There is no indication of outliers or other effects indicating special causes.

Obviously, potentially important sources of variation are overlooked in this case. A more complete account of sources of variation in this example includes:

- Random variation in thickness at a fixed position under the grindstone within the same batch;
- (Systematic) variation between positions within the same batch;
- Random variation between different batches.

Table 7.2: Thickness measurements (μm).

Batch	Position 1	2	18	19	28	Batch	Position 1	2	18	19	28
1	240	243	250	253	248	16	237	239	242	247	245
2	238	242	245	251	247	17	242	244	246	251	248
3	239	242	246	250	248	18	243	245	247	252	249
4	235	237	246	249	246	19	243	245	248	251	250
5	240	241	246	247	249	20	244	246	246	250	246
6	240	243	244	248	245	21	241	239	244	250	246
7	240	243	244	249	242	22	242	245	248	251	249
8	245	250	250	247	248	23	242	245	248	243	246
9	238	240	245	248	246	24	241	244	245	249	247
10	240	242	246	249	248	25	236	239	241	246	242
11	240	243	246	250	248	26	243	246	247	252	247
12	241	245	243	247	245	27	241	243	245	248	246
13	247	245	255	250	249	28	239	240	242	243	244
14	237	239	243	247	246	29	239	240	250	252	250
15	242	244	245	248	245	30	241	243	249	255	253

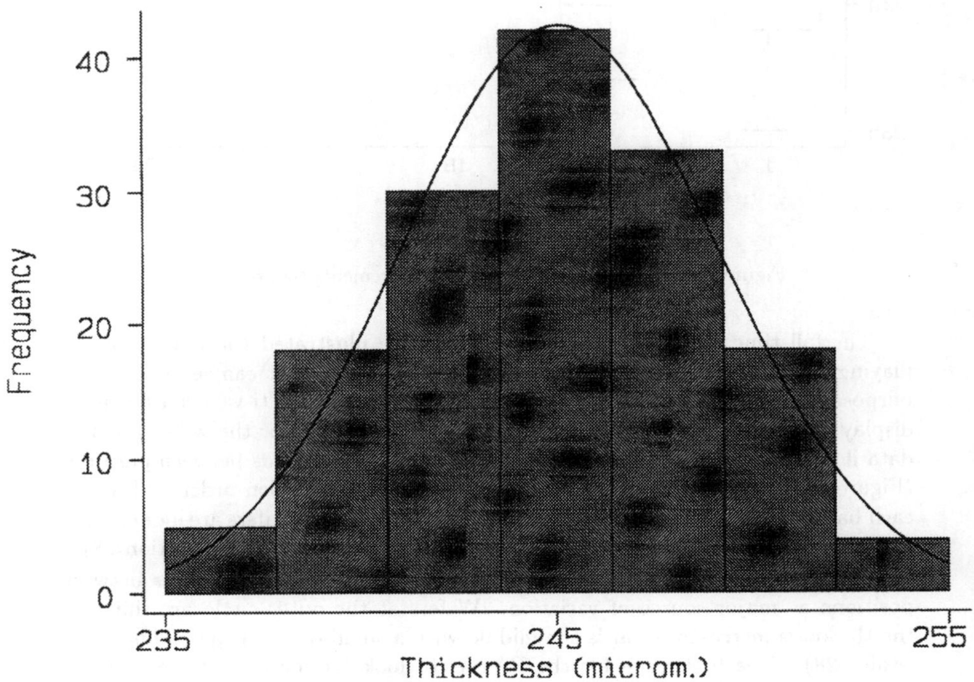

Figure 7.5: Histogram of all thickness measurements.

These components can be assessed graphically and by more advanced statistical analyses. ***Box plots*** per position (Figure 7.6) clearly illustrate substantial systematic differences between positions. The order of magnitude is about 8 μm between largest and smallest median. The ***interquartile range*** within a position is about 5 μm. Moreover, outliers are indicated (positions 1, 18 and 28). Note that the variation within one position includes variation between batches as well as variation within a batch at the same position.

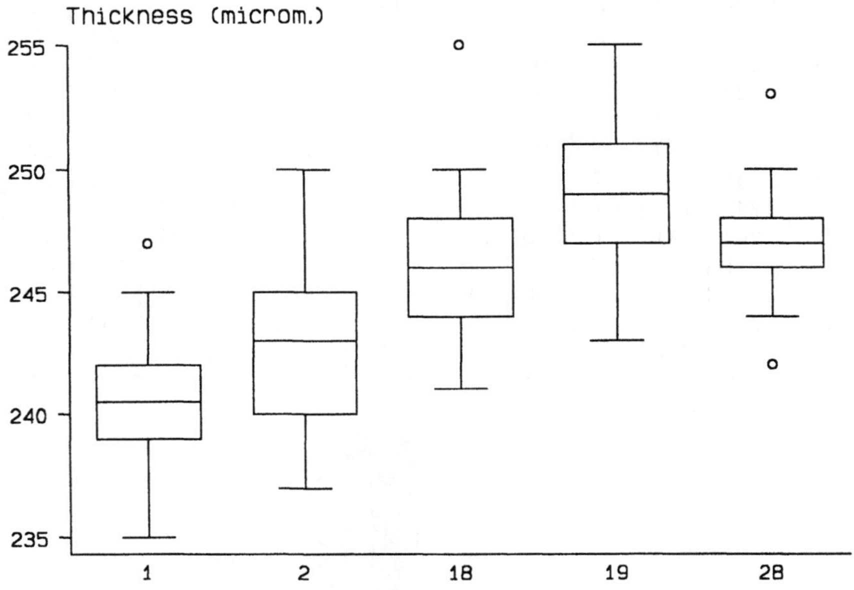

Figure 7.6: Box plots of thickness measurements by position.

The fall time data from the previous example illustrated the impact of displaying the data over time. In this case, the multi-vari chart can serve the same purpose (Seder (1990) and section 9.2 of this book). The multi-vari chart aims to display simultaneously different components of variation. For the wafer grinding data it seems worthwhile to display variation in time as well as between positions (Figure 7.7). In one chart the batches are shown in production order and within each batch the individual results per position are indicated (wafers are indexed as 1 (position 1), 2 (position 2), 3 (position 18), 4 (position 19) and 5 (position 28)).

The chart clearly identifies the systematic differences within batches between positions as major source of variation. Wafers at the outer circle are thinnest; the thickness increases towards the middle with a small drop on the middle circle (wafer 28). The first issue for the PAT is to look for methods to reduce this systematic difference, which is caused by the fact that grindstone and grinding table are not exactly parallel. The wafer grinding PAT was able to reduce this difference by improving the periodic maintenance procedures.

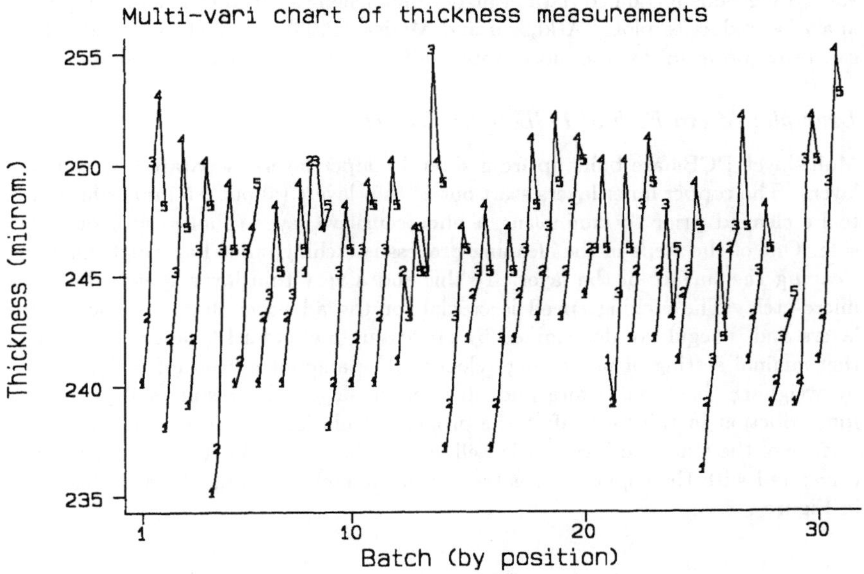

Figure 7.7: Multi-vari chart of thickness measurements.

The graphical analysis of the data is confirmed by an analysis of variance, including batch and position as factors (Table 7.3).

Table 7.3: Analysis of variance of thickness measurements.

Source	Sum of Squares	df	Mean Square	F ratio
Batch	601	29	20.7	5.92
Position	1218	4	354.4	101.20
Residual	406	116	3.5	
Total*	2425	149		

* Corrected for the mean.

7·3.3 Exploration in More Dimensions

In many cases a multitude of continuous and possibly correlated measurements is collected. In such cases outliers and patterns in the data have to be explored in more dimensions. In this context methods developed for multivariate Statistical Process Control can be appropriate as explorative tools. Recent reviews on the use of multivariate control charts can be found in Wierda (1994) and Mason *et al.* (1997). As a first screening of the data, well chosen scatter diagrams will provide insight into dependence and possible outliers or other special cause effects. Advanced methods for joined data of higher dimension (such as principal component analysis, stalactite plots (Atkinson and Mulira (1993) and Sullivan *et al.*(1995)) can prove powerful for the more statistically educated quality professional.

Example: Micro-Etch of PCB Inner Layers

Multi-layer PCBs are built up from several copper layers, separated by isolation layers. The copper inner layers start out as thin layers (about 0.5 mm) which have to be cleaned prior to laminating a photosensitive layer (photoresist) on either side. One of the steps in the cleaning process is etching, aimed at roughening and cleaning the surface of the layer. To this end a very thin layer is etched off, the micro-etch. The etching speed is crucial for the adhesive characteristics of the layers and is regularly determined by processing a standard type copper layer at the nominal setting of processing velocity. The amount etched off is determined by weighing the layer before and after processing. The result is expressed as μm reduction in thickness after the proper calculations based on the size of the surface of the standard layer. It is well known that the etching speed is positively correlated with the copper concentration in the etching bath, which is illustrated in Figure 7.8.

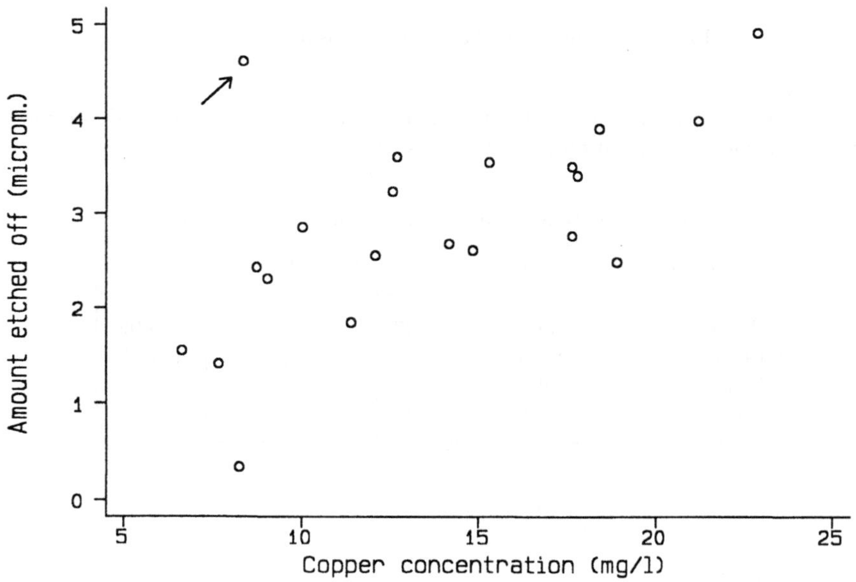

Figure 7.8: Positive correlation between amount etched off and copper concentration.

There is obviously one outlier: a special cause present! This would not have been detected by evaluating the amount etched off and copper concentration separately. In this case a measurement error might be the cause. The weighing involved is quite sensitive to outside disturbances. To circumvent such problems and improve process control in general, a feedback control mechanism based on the (copper) colour of the etching fluid was installed. This significantly reduced variation in copper concentration and in the adhesive characteristics of the layers.

7·3.4 Exploring Counts

Thus far we concentrated on continuous numerical process data. Occasionally we have to rely on nominal data: the number of pads on a PCB with insufficient solderability, for example. The concept of common cause and special cause variation still remains applicable, but now to the counts or to the fraction of non-conforming products in a sample. A primary distinction in counting errors or non-conforming items is:

1 Counts of errors per (large) unit;

2 Counts of separately manufactured distinct products or parts.

The count of the pads with insufficient solderability is an example of the first type, considering the PCB as (large) unit. Such counts may display a systematic distribution over the unit, quite similar to the systematic differences in thickness within a batch in the grinding wafers example.

If a PCB is considered either as conforming or non-conforming depending on the seriousness of tin veils over the PCB, the number of non-conforming PCBs in a sample is likely to be the count of interest. This is an example of the second type of counts, with PCBs as the separately manufactured distinct products.

The distinction between the two types of counts is reflected in the way common cause variation is investigated.

Example: PCB Production

Solderability of pads on a PCB is (partly) determined by the performance of Hot Air Levelling (section 7.2). In this process step the PCBs are treated vertically, which makes it plausible that there will be differences between the upper and lower part of a PCB in solderability. The count of non-conforming pads per PCB should thus be split up in the count over the upper part and the count over the lower part of the PCB. Results over 20 PCBs of the same production type analysed this way are presented in Figure 7.9. This clearly shows a systematic difference: pads on the lower part tend to have more solderability prlblems than the ones in the upper part.

As before, actions on the process have to be taken to improve the situation. In this case adjusting the angles of the high pressure air flow, temperature, and processing time were investigated to optimise the process. ■

Thus it is important to investigate systematic effects when dealing with counts per unit. It should be noted that a presence of systematic effects violates the first of the following essential assumptions usually made for counts of faults per unit:

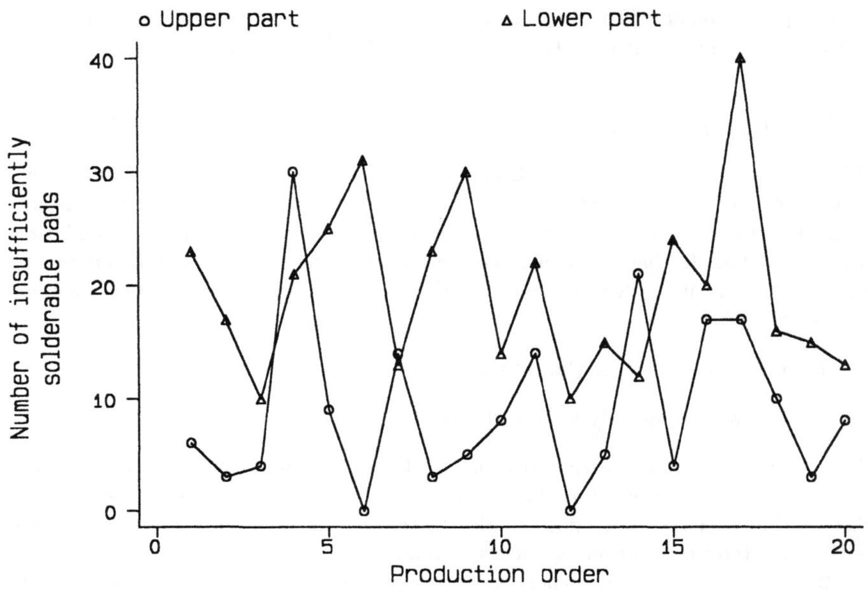

Figure 7.9: Trend chart counts of pads with non-conforming solderability.

- Faults occur randomly across the unit: the count of faults in a chosen subarea is independent from the count of faults in another disjoint area;
- The mean number of faults per unit is constant, when the process is under control.

The second assumption is violated in the case where the number of faults per unit is subject to extra random variation between units. This is also not uncommon in real life applications.

Counts of separately processed distinct products are difficult to evaluate for systematic effects, unless very large sample sizes prove possible and economically sensible. Instead of counts it is common to evaluate fractions of non-conforming within the sample. The usual assumptions underlying such fractions are:

- Every product has the same probability p of ending up non-conforming;
- Whether a specific product becomes non-conforming, does not depend on the other products;
- The probability p remains constant when the process is under control.

For the same reasons mentioned for counts per unit, the latter two assumptions are not always met in practice.

As a consequence the standard control charts for counts (c-chart) and fractions (p-chart) are not generally applicable. Robust alternatives are provided in the next chapter.

7·4 Sampling Plan and Process Characteristics

7·4.1 Sampling Plans

In Shewhart's era, production processes were not nearly as complex as nowadays. Most products were manufactured and assembled in large series, in relatively comprehensive production processes. Sampling plans for such processes were determined by establishing a *sample frequency* (*e.g.*, per hour, per shift, per day) and prescribing how products should be sampled within each period (completely at random over the total period, around fixed time points, series of consecutively produced products). Sampling for control charts was based on the concept of *rational subgroups*: sample frequency and product selection within a sample were to be such that:

1 Variation between products within a sample includes all common cause variation;

2 Special causes occur between samples.

All traditional methods of setting control chart limits are based on this leading principle (see Montgomery (1996)).

Contemporary manufacturing processes are considerably more complex, with a large diversity of products and high technology processing. Sampling plans for control charts as part of the SPC control loop need to take this complexity into account to be effective. Thus adequate sampling plans are based on the statistical analysis of the data gathered in the diagnostic phase and the engineering knowledge of the process at hand. The components of variation identified and explained according to this knowledge form the basis of *rational sampling* (Palm, 1992): sampling that is purposely not random and aims to describe the important variance components. The components can roughly be distinguished in systematic (fixed) effects and random effects.

Systematic effects arise when one or more factors (such as grindstone and grinding table not being parallel) affect the quality characteristics of interest consistently over time. It may be possible to remove or reduce the cause for such systematic effects; this should first be attempted. Only to the extent that this is not possible (or economically sensible) such systematic effects are part of the common cause variation. The systematic effects should then also be controlled through appropriate control charts. Typical examples include:

□ Differences between machines, designed to operate identical;

□ Different positions on a mould;

□ Different leads from the same equipment;

□ Tool wear.

Random effects are responsible for variation occurring randomly over time which cannot be traced back to specific causes for individual products or measurements. The following sources of random variation can be distinguished:

□ Measurement variation;

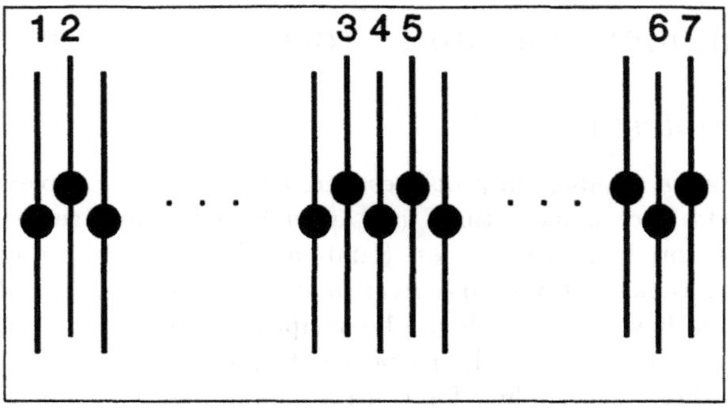

Figure 7.10: Holder with 48 medium power rectifiers after glass deposition.

- Variation between products manufactured under the same conditions;
- Variation between batches or samples of products;
- Variation between shifts, days, series *etc.*.

Random variation is also causally linked to process, product, machine, man and environmental factors. However, many factors contribute to the total random variation and each of these factors only has a relatively small impact. Thus the causes cannot be traced on an individual basis.

In many applications the ***nested*** structure of random variation as indicated by the previous list is encountered. Quantification of each of the components—by means of analysis of variance—can prioritise the components to be reduced. In most cases, reduction of random variation is more difficult and will generally require system changes (machine, environment, materials).

Example: Glass Deposition for Medium Power Rectifiers

A medium power rectifier is a special type of diode, and one manufactured by Philips Semiconductors in Stadskanaal. The process step to be considered is 'glass deposition'. To passivate the crystal (the heart of the diode, made of semiconductor material), a liquid glass bead is deposited. The glass is deposited simultaneously on 48 products in a holder (Figure 7.10). After glass deposition this holder is carried through to a drying position. The process parameter used to monitor the process is the diameter of the bead. It is known from engineering considerations that there is a systematic drift in diameter across the length of the holder which cannot be easily corrected. It is owed to processing the holder in the lateral direction. Thus sampling for process control was designed as follows: each shift one holder is sampled, and the diameters of 7 products are measured. These products are systematically selected (see Figure 7.10), two from the left hand side of the holder, three from the middle, and two from the right hand side.

The systematic drift over the holder can now be monitored (see below) because of the non-random sample covering the holder. Random variation between beads of different crystals can be evaluated, owing to two or three measurements coming from beads very close together. The systematic drift can be neglected at such close distance. ■

Silver evaporation and grinding wafers (discussed previously) were examples of similar sampling schemes. A general recipe for constructing ***rational sampling*** schemes cannot be given. The particular process conditions, engineering knowledge and relative size of random variation components will have to be taken into account and may lead to very different schemes in seemingly similar situations. In first implementations, an industrial statistician can be consulted to assist in setting up useful sampling schemes. From the discussions in Chapter 6 it has become clear that samples sizes (frequency) should be sufficiently large, particularly for the initial stage when control charts need to be set up. The need to assess both systematic effects and nested random variation components further emphasises this finding.

7·4.2 Process Characteristics to Be Controlled

The sampling scheme and identified (nested) components of variation lead to four categories relevant for control charts: systematic effects (yes/no) combined with extra random variation between samples (yes/no). This is summarised in Table 7.4.

Table 7.4: Variation, sampling schemes and characteristics to be controlled.

		Random variation (common cause)	
		Within Samples Only	Within and between Samples
Systematic Effects	None (§8.3)	I Mean and range of sample	II Mean and range of sample; adjusted control limits
	Within Samples (§8.4)	III Mean, within-sample differences and mean range of close measurements	IV Mean, within-sample differences and mean range of close measurements; adjusted control limits

Situation I corresponds to the ***rational subgroup*** concept and is treated as standard situation in most textbooks. For continuous measurements this leads to the well known Shewhart type of $\overline{X}$-R chart. Unfortunately this situation is rarely encountered in practice.

Chemical batches are a typical example of ***situation II***. The processing of a single batch (mixing, heating, blending, *etc.*) is such that variation within a batch can be considered random. Since raw materials and processing conditions (heat, pressure, ...) operate at the batch level there are many factors causing extra between batch variation. Thus series of samples within batches will display both kinds of random variation. Control chart limits need to take account of both components, since they are generally beyond the control of the shop floor operator running the process.

Production of continuous series of the same type of produce can lead to *situation III* if the same type measurements are repeatedly performed on the same product. Examples include: measurement of tin layer thickness at different positions on a PCB after Hot Air Levelling, and the diameter of cylindrical bearings at different positions. The systematic effects can than adequately be monitored by charting the sample means of well chosen differences within products.

Situation IV actually combines batch types of phenomena with the systematic effects from situation III. In our view this should be referred to as the *standard situation*. The example below describes this standard situation and indicates the type of characteristics to be charted.

Example (continued): Glass Deposition for Medium Power Rectifiers

The sample that describes the systematic drift over the t-th holder consists of seven measurements: (X_{t1}, X_{t2}) left, (X_{t3}, X_{t4}, X_{t5}) middle and (X_{t6}, X_{t7}) right (Figure 7.10). Of course, the sample mean:

$$\overline{X}_t = \tfrac{1}{7}(X_{t1} + X_{t2} + X_{t3} + x_{t4} + x_{t5} + X_{t6} + x_{t7}),$$

can be charted to control the mean glass bead diameter over time. The systematic drift over a holder can be split into two components: linear and quadratic (see Figure 7.11).

$$\begin{aligned} D_{Lt} &= \tfrac{1}{2}(X_{t1} + X_{t2}) - \tfrac{1}{2}(X_{t6} + X_{t7}), \\ D_{Qt} &= \tfrac{1}{4}(X_{t1} + X_{t2} + X_{t6} + X_{t7}) - \tfrac{1}{3}(X_{t3} + X_{t4} + X_{t5}). \end{aligned}$$

The random variation between bead diameters at (nearly) the same position can be controlled by adding the three ranges of the three groups of close beads:

$$R_t = \text{range}(X_{t1}, X_{t2}) + \text{range}(X_{t3}, X_{t4}, X_{t5}) + \text{range}(X_{t6}, X_{T_7}).$$

Many other possibilities exist for this characteristic, but we chose to keep in close agreement to the $\overline{X}$-R method.

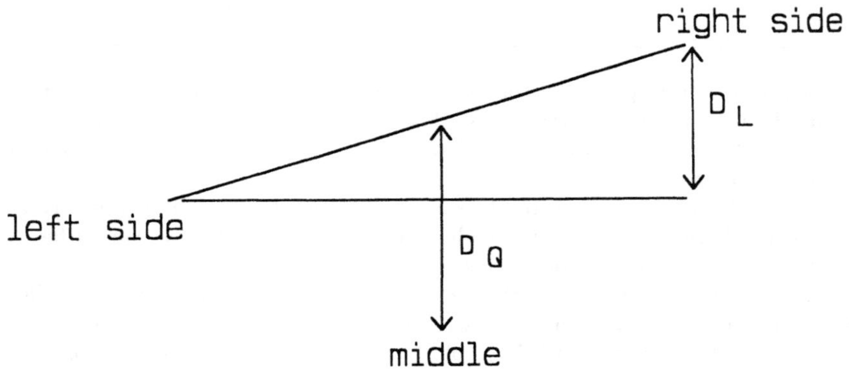

Figure 7.11: Systematic drift in bead diameters after glass deposition.

The choice of the linear and quadratic contrasts D_{Lt} and D_{Qt} also resulted from a principal component analysis of data from the process (Does, Roes and Trip (1995)). Moreover, these components could be understood in physical terms, and we were able to provide an OCAP for each component.

7·5 Control Chart Layout

Once the characteristics to chart have been determined, control chart limits (LCL, CL and UCL for Shewhart types of charts) need to be established. This statistically oriented subject is dealt with in detail in the next chapter. Actual layout of the chart as visible for the operator on the shop floor also needs attention. Despite the current advanced computer age, charts with pencil and paper may occasionally have an important function. Layout of charts is the subject of this section; the principles readily translate to computer or paper and pencil implementations of control charts.

The use of control charts on the shop floor is aimed at registering and interpreting important information on the process as it is running currently. A comprehensive layout and sufficient room for notes are important, as are clearly visible display and easy access at the shop floor. On the control chart form we distinguish the following elements:

□ Basic information on the process step it is applied to;

- Information on type of measurement, measurement method and frequency;
- Adequate room for the raw data;
- The actual charts;
- References to OCAPs, logs and Standard Operating Procedures.

The information thus displayed should be easy accessible not only to operators, but also to management and customers.

Information on the process step would typically contain the name and purpose of the process step, department and equipment used. Similarly, measurement equipment, frequency, and sample sizes are indicated. The individual raw data directly displayed serve as a valuable aid in investigating out of control situations. The actual charts are grouped and justified vertically (with the raw data also), to identify out of control conditions in one look. Highlighting areas with colours (*e.g.*, red outside the LCL and UCL) will further increase effectiveness. As an example a control chart from Philips Stadskanaal is displayed in Figure 7.12.

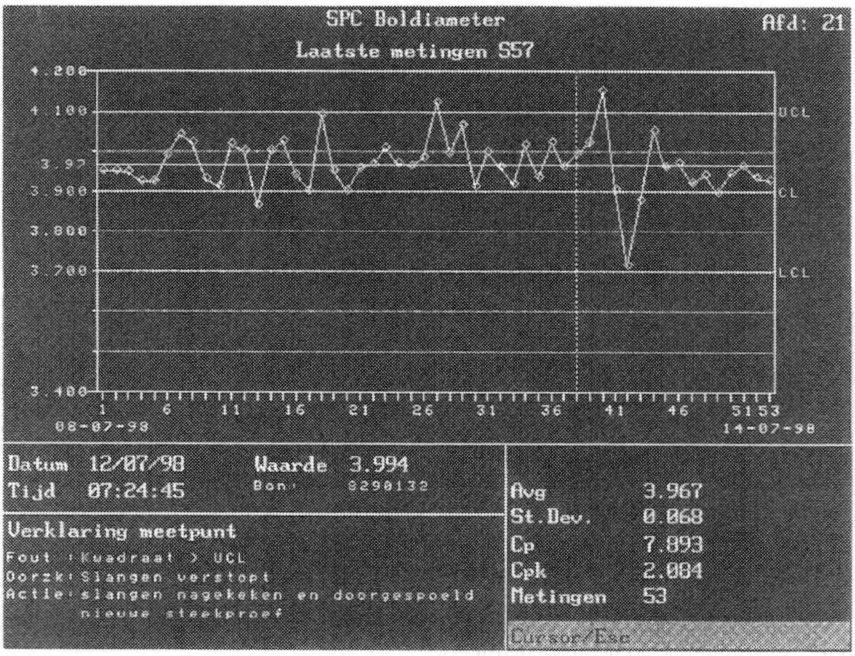

Figure 7.12: Example of a complete control chart.

Chapter 8

Control Charts

8·1 Introduction

In the previous chapter we focused on the analysis of process data to identify and quantify components of variation inherent to the process. This analysis formed the basis for selecting the (derived) characteristics to be monitored by means of control charts. The principal layout of the control chart was also discussed, which leaves one important subject to cover: how to calculate appropriate control chart limits? This chapter covers this issue for a range of Shewhart type of control charts. 'Traditional' charts, such as the $\overline{X}$-R, as well as the charts applicable to fixed effects within samples are described and discussed. Section 8.2 deals with the chart for individual measurements. This type of chart is important for monitoring typical process parameters—such as temperature, pressure and concentrations—but the methodology is also applied in more complex situations to provide robust control limits. Next, the traditional $\overline{X}$-R chart is briefly introduced and the extension necessary for the situation with extra between sample variation is covered. Section 8.4 provides a more in depth treatment of control charts in the 'standard situation' of both extra between sample variation and within sample fixed effects.

All these charts rely on a sufficient amount of process data being available before the control chart is implemented as part of the SPC control loop. This will, in general, be possible for high production volumes (albeit with a great diversity of products). In current manufacturing practices true low volume manufacturing is encountered. The production of wafer steppers by ASM Lithography at a volume of about 200 per year provides an illustrative example. This requires adapted control chart methods, which can be implemented after a (very) small amount of process data has been collected (section 8.5).

Finally, control charts for counts and fraction are described and the concept of pre-control is discussed in comparison with the control chart approach for monitoring processes.

8·2 Control Charts for Individual Measurements

8·2.1 Estimating Mean and Dispersion for Control Chart Limits

Process parameters such as temperature, pressure and processing time tend to be evaluated for stability at every individual measurement which comes available. Also for other characteristics situations arise for which monitoring will be based on individual measurements, *i.e.*, , sample size $n = 1$. A recent treatment of general aspects of control charts for individual measurements can be found in Roes *et al.*(1993). The first k, say, of the consecutive measurements $X_1, X_2, X_3, \ldots,$ form the initial sample on which control chart limits have to be based. Under the assumption that the X_t are distributed according to a normal distribution with mean μ and standard deviation σ, the calculation of control limits requires estimation of μ and σ based on $(X_1, X_2, X_3, \ldots, X_k)$. The estimators of μ that reasonably can be used are the sample mean $\overline{X}$ and the sample median Med(X). The real issue is of course the choice of the estimator of σ, and a number of estimators suitable for control charting are briefly discussed in this section. The most efficient estimator for σ is the ***standard deviation***:

$$S = \sqrt{\frac{1}{k-1}\sum_{t=1}^{k}(X_t - \overline{X})^2} \ . \tag{8.1}$$

It is well known that S^2 is distributed as $\sigma^2/(k-1)$ times a χ-squared distributed random variable with $k-1$ degrees of freedom (denoted by χ^2_{k-1}). As a consequence S is distributed as $\sigma/\sqrt{k-1}$ times the square root of a variable with the χ^2_{k-1} distribution. An unbiased estimator of σ can be obtained by dividing S by $c_4(k)$ (these $c_4(k)$ are tabulated by Duncan (1986)). In setting control limits other estimators of σ may be preferred for various reasons. A first point in this respect is that S is sensitive to trends and shifts in the X_t's over time. This can be a particular disadvantage when setting up control charts for past data, that might not be under control. As a consequence estimators have been considered that do take the time order into account, and are possibly less sensitive to these deviations. Kamat (1953) discusses the ***mean successive difference***:

$$MR = \frac{1}{k-1}\sum_{t=2}^{k}|X_t - X_{t-1}|, \tag{8.2}$$

which is also known as the ***mean of the moving ranges***, hence the abbreviation. MR is biased; division by $d_2(2) = 1.128$ (see Duncan (1986)) generates an unbiased estimator. MR is much less influenced by a trend in the mean of the observations than is the standard deviation. This is also confirmed by the simulations performed by Harding, Lee and Mullins (1992).

A more general resistant measure of variation is given by the ***inter-quartile range***:

$$IQR = X_{k-[k/4]+1:k} - X_{[k/4]:k} \tag{8.3}$$

where [,] denotes the integer part and $X_{i:k}$ denotes the i-th order statistic. This estimator of the standard deviation actually discards 50% of the observations, which makes it particularly robust against outliers. $IQR/1.349$ provides an approximately unbiased estimator of the standard deviation.

Control chart limits—based on 3σ principles—can be set as follows:

$$\begin{aligned} \text{UCL} &= \overline{X} + 3V, \\ \text{CL} &= \overline{X}, \\ \text{LCL} &= \overline{X} - 3V. \end{aligned} \tag{8.4}$$

In these equations V can be substituted by $S/c_4(k)$, $MR/1.1.28$ or $IQR/1.349$ and $\overline{X}$ can be replaced by the median of the observations, Med(X). The use of $\overline{X}$ and $S/c_4(k)$ leads to the chart already introduced in Chapter 6.

Example (continued): Reticle Handling Elevators

In section 7.3 we mentioned the reticle handling elevators mounted in wafer steppers of ASM Lithography. It is clear from the type of process and total production volume that evaluation after every mounted elevator is important to control the process. The actual measurement to evaluate proper mounting is the so called fall time, the time (in (s)) it takes a defined weight to move the arm of an elevator from fully up to fully down. The first 20 measurements are displayed in Table 8.1, along with basic calculations for the sample mean and MR.

Table 8.1: Fall time measurements (s), sample mean and mean moving range (MR).

Elevator	Time (s)	Absolute Difference	Elevator	Time (s)	Absolute Difference
1	2.12	–	11	3.69	1.34
2	1.81	0.31	12	2.10	1.59
3	1.93	0.12	13	2.05	0.05
4	3.46	1.53	14	2.60	0.55
5	3.50	0.04	15	1.88	0.72
6	2.25	1.25	16	2.28	0.40
7	2.44	0.19	17	2.90	0.62
8	2.85	0.41	18	1.40	1.50
9	1.75	1.10	19	1.90	0.50
10	2.35	0.60	20	3.60	1.70
		Mean		2.44	0.76

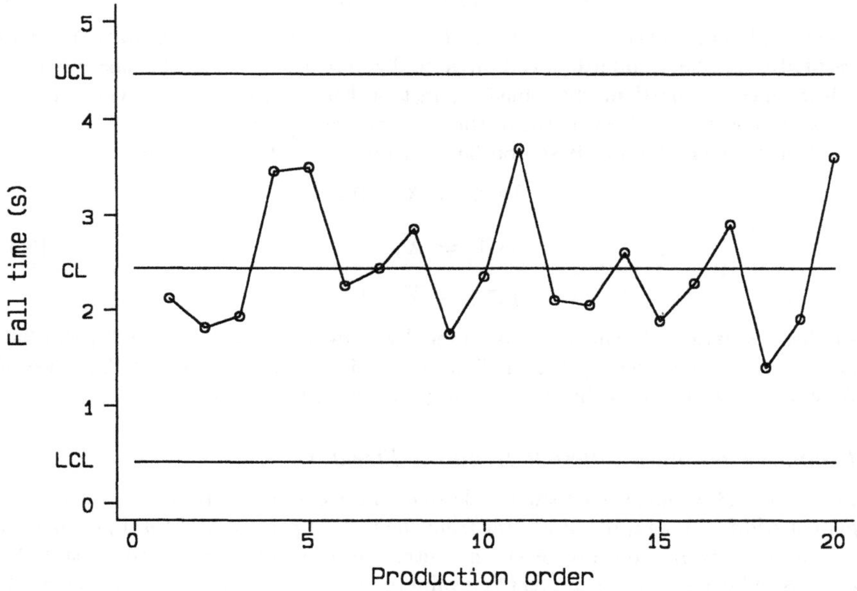

Figure 8.1: Control chart for fall times of reticle handling elevators.

The control chart limits are (LCL, CL, UCL) = $(2.44 - 2.66 \times 0.76, 2.44, 2.44 + 2.66 \times 0.76 = (0.42, 2.44, 4.46)$, which produces the control chart of Figure 8.1. It is clear that calculations are simple enough to be implemented in a spreadsheet.

8·2.2 *Efficiency of the Proposed Charts*

A general recommendation for the choice between the control chart options is partly based on comparing the variances of the respective estimators of spread. The (intuitive) basis for this is that this variance determines the accuracy with which the limits are set at a given fixed sample size. In fact, if the variance of estimator V_1 is twice that of estimator V_2, the sample size needed with estimator V_1 should be twice as large as that with V_2 to achieve the same accuracy in setting limits. Since it established that $S/c_4(k)$ is the most efficient unbiased estimator, we compare the variances of the other estimators with $\text{var}(S/c_4(k))$:

$$R_{\text{eff}}(S, V) = \frac{\text{var}\left(\dfrac{S}{c_4}\right)}{\text{var}(V)} \times 100 \tag{8.5}$$

the customary efficiency measure, of which the reciprocal was used in Cryer and Ryan (1990) for $S/c_4(k)$ and $MR/d_2(2)$.

The results indicate (Table 8.2) that MR has a possibly acceptable efficiency, but the efficiency loss for IQR is dramatic. Hence IQR would provide rather unreliable limits compared to the other estimators.

Table 8.2: Efficiency comparison for $k = 30$ and $k = 60$.

Chart	Efficiency (%)	
	$K = 30$	$k = 60$
$\overline{X}, S$	100	100
$\overline{X}, MR$	62	61
$\overline{X}, IQR$	38	38

In addition to efficiency, robustness to undetected special causes in the initial sample is an important consideration for recommendation in routine practice. In this respect the sample standard deviation S is very sensitive to shifts, trends, and outliers. MR provides a better alternative, as mentioned before. Thus we recommend the individual measurements chart based on MR as the general method for routine practical applications.

8·2.3 Chart the Measure of Spread

The issue of displaying a chart of some measure associated with the spread in the X_t (*e.g.*, a chart of $MR_t = |X_t - X_{t-1}|$ is a matter of some controversy. Nelson (1982) advises not to chart the MR_t, and Wheeler and Chambers (1992) stress the indispensability of such a chart. Apart from noting that the moving ranges are correlated, Nelson states that "...the chart of the individual observations actually contains all the information available". The consequences of this for the charting procedure becomes particularly clear if we evaluate the counter argument of Wheeler and Chambers on probabilistic grounds. They argue (p. 221) "It should also be noted that it is possible for the moving range to be out of control even when the two individual values are within their 'control limits'. For this reason one should *always* plot the moving ranges on a control chart." We share Nelson's view and arguments, and give a probabilistic justification. To elaborate, consider the conditional probability:

$$P\{\text{Signal on } MR_t \mid X_{t-1}, X_t \text{ under control}\} \tag{8.6}$$

in case where the X_{t-1} are normally distributed with known μ and σ. This probability was evaluated for X_{t-1} and X_t identically distributed (the 'real' under control situation), and for out of control situations in X_t described by either a jump in mean level of size $a\sigma$ for $a = 0.5$ to 3.0, or an increase of σ to $b\sigma$ for $b = 1.5$ to 3. The conditional probability can be compared with the unconditional probability of signalling the out of control situation on the control chart for the X_t, to judge the contribution the moving range chart has to the power of discovering an out of control situation when it actually occurs. The results are summarized in Table 8.3.

Table 8.3: Conditional probability of signalling on MR chart and probability of signalling on X-chart.

Jump a in Mean	$P\{\text{Signal on } MR_t \mid X_{t-1}, X_t \text{ under control}\}$	$P\{\text{Signal on } X_t\}$
0	0.000 58	0.0020
1.0	0.002 28	0.0183
2.0	0.008 11	0.1378
3.0	0.016 59	0.4641
Increase to $b\sigma$		
1.0	0.000 58	0.0020
2.0	0.004 50	0.1223
3.0	0.005 96	0.3030

It is obvious from Table 8.3 that the probability of signalling an out of control situation only through the chart of MR_t is indeed very small. These power considerations were recently confirmed in a more extensive study by Amin and Ethridge (1998). Added to the difficulties in interpreting the MR chart owed to the serial correlation, it is clear that displaying a MR chart can confuse a straightforward interpretation. It is much better understood to use supplementary run rules (Chapter 6, and Nelson (1984) and Does and Schriever (1992)), if the intention is to increase the power of the control chart.

In summary, we recommend the individual X chart with limits based on the mean moving range MR, without displaying the additional 'moving range chart'. This will ensure reliable control charts if the initial sample size is about 50 measurements or larger. If the time required to collect such a sample is prohibitively long, low volume control charts should seriously be considered (section 8.5).

8·3 Control Charts for Samples without Systematic Effects

Standard textbooks usually assume samples for control charts which are only subject to within sample random variation. This corresponds to situation I from Table 7.4. Processes which in addition are subject to extra between sample random variation, such as batch process, are similar with respect to control charting (situation II). Both situations are dealt with in this section, based on the sample measurements as schematically displayed in Table 8.4.

Table 8.4: Sample measurements.

Sample	Measurements	Mean	Range
1	$X_{11}, X_{12}, X_{13}, \ldots, X_{1n}$	$\overline{X}_1$	R_1
2	$X_{21}, X_{22}, X_{23}, \ldots, X_{2n}$	$\overline{X}_2$	R_2
3	$X_{31}, X_{32}, X_{33}, \ldots, X_{3n}$	$\overline{X}_3$	R_3
$\ldots t \ldots$	$X_{t1}, X_{t2}, X_{t3}, \ldots, X_{tn}$	$\overline{X}_t$	R_t
k	$X_{k1}, X_{k2}, X_{k3}, \ldots, X_{kn}$	$\overline{X}_k$	R_k

8·3.1 Situation I

All measurements are assumed to arise from the same probability distribution if the process is under control, hence the following model holds for measurements X_{tj}:

$$X_{tj} = \mu + E_{tj} \quad \text{for } j = 1, 2, \ldots, n \text{ and } t = 1, 2, 3, \ldots, \tag{8.7}$$

where μ indicates the constant overall mean and E_{tj} models the random variation. The E_{tj} are assumed to be mutually independent and normally distributed with means 0 and variance σ_e^2. Thus the sample means $\overline{X}_t$ are also normally distributed with mean μ and standard deviation $\sigma_e/\sqrt{n}$. If μ and σ_e are known, control chart limits for the sample means are simply:

$$\begin{aligned} \text{UCL} &= \mu + \frac{3\sigma_3}{\sqrt{n}}, \\ \text{CL} &= \mu, \\ \text{LCL} &= \mu - \frac{3\sigma_e}{\sqrt{n}}. \end{aligned}$$

In Chapter 6 one of the standard options in case parameters have to be estimated was introduced: $\overline{X}$-S. Here we concentrate on the mean of the sample ranges $R_t = \text{Max}(X_{t_1}, X_{t_2}, X_{t_3}, \ldots, X_{tn}) - \text{Min}(X_{t_1}, X_{t_2}, X_{t_3}, \ldots, X_{tn})$ as estimator of σ_e. Hence let:

$$R = \frac{1}{k}\sum_{t=1}^{k} R_t. \tag{8.8}$$

Then $\overline{R}/(\sqrt{nd_2(n)})$ is an unbiased estimator of $\sigma/\sqrt{n}$. Thus the control limits based on k samples of size n become:

$$\text{UCL}_{\overline{X}} = \text{CL}_{\overline{X}} + 3\,\frac{\overline{R}}{\sqrt{nd_2(n)}} = \text{CL}_{\overline{X}} + A_2\overline{R},$$

$$\text{CL}_{\overline{X}} = \frac{1}{k}\sum_{t=1}^{k}\overline{X}_t = \overline{\overline{X}}, \tag{8.9}$$

$$\mathrm{LCL}_{\overline{X}} = \mathrm{CL}_{\overline{X}} - d\frac{\overline{R}}{\sqrt{n d_2(n)}} = \mathrm{CL}_{\overline{X}} - A_2\overline{R}.$$

Using similar arguments based on the model (8.7), control limits for the sample ranges can be determined as:

$$\begin{aligned} \mathrm{UCL}_R &= D_4\overline{R}, \\ \mathrm{CL}_R &= \overline{R}, \\ \mathrm{LCL}_R &= D_3\overline{R}. \end{aligned} \tag{8.10}$$

The constants A_2, D_3, and D_4 all depend on the sample size, and are tabulated in most textbooks on statistical quality control, such as Duncan (1986) and Montgomery (1996). A selected range is presented in Table 8.5.

Table 8.5: Control chart constants for basic model (8.7).

Sample	Chart for $\overline{X}_t$	R_t	
(n)	A_2	D_3	D_4
2	1.880	0	3.267
3	1.023	0	2.575
4	0.729	0	2.282
5	0.577	0	2.115
6	0.483	0	2.004
7	0.419	0.076	1.924
8	0.373	0.136	1.864
9	0.337	0.184	1.816
10	0.308	0.223	1.777

Despite the $\overline{X}$-R chart being very well known and prominently dealt with in many standard textbooks, we have not encountered real life processes to which it could be appropriately applied. The following example is therefore to a great extent fictitious.

Example: Silencer for Car Door (Slamming)

In a specific type of (European) car the sound of slamming car doors is silenced by applying a plastic cylinder around the pin that catches the car door lock. These hollow cylinders can be either too wide, resulting in premature wear, or too small, which complicates assembly. Tight control of the inner cylinder diameter is therefore of the utmost importance. In the silencer production line 5 cylinders are sampled and measured every hour. Results (sample means and ranges) of 20 of such samples are presented in Table 8.6.

The overall mean is 9.507 mm and the mean range 0.229 mm. Thus from equations (8.8) the control limits for the sample means are established as ($\mathrm{LCL}_{\overline{X}}, \mathrm{CL}_{\overline{X}}, \mathrm{UCL}_{\overline{X}}$) = (9.375, 9.507, 9.639); the control limits for the sample ranges (8.9): ($\mathrm{LCL}_R, \mathrm{UCL}_R$) = (0, 0.484). The corresponding charts are displayed in Figure 8.2.

Table 8.6: Means and ranges of 20 samples of 5 inner cylinder diameters (mm).

t	$\overline{X}_t$	R_t	t	$\overline{X}_t$	R_t	t	$\overline{X}_t$	R_t	t	$\overline{X}_t$	R_t
1	9.487	1.884	6	9.548	0.242	11	9.513	0.181	16	9.463	0.238
2	9.495	0.267	7	9.479	0.365	12	9.453	0.224	17	9.498	0.280
3	9.462	0.373	8	9.506	0.129	13	9.575	0.195	18	9.533	0.102
4	9.558	0.106	9	9.521	0.299	14	9.553	0.188	19	9.493	0.202
5	9.496	0.301	10	9.504	0.223	15	9.504	0.284	20	9.502	0.195

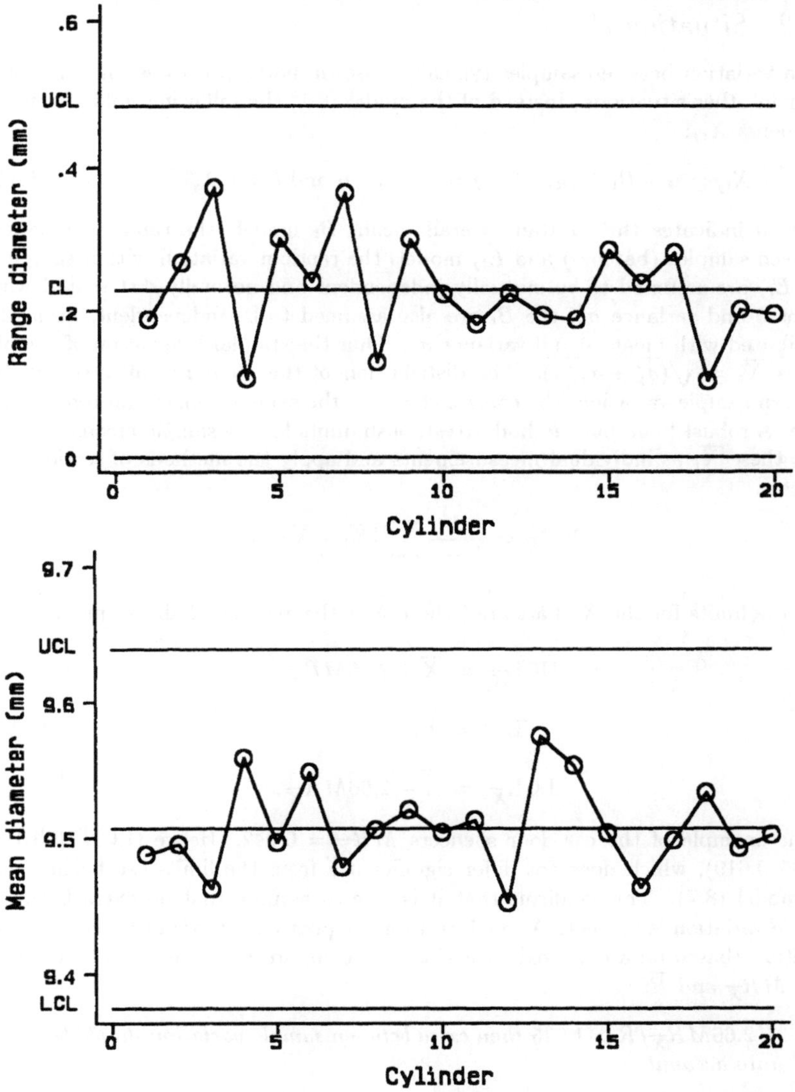

Figure 8.2: Control charts for sample ranges and sample means of cylinder diameters.

The charts will generally be evaluated jointly. It should be noted, however, that a special cause on the R-chart will generally be caused by outlying measurement(s) within one sample. The sample mean for that particular sample will then probably also be affected, if sample sizes are small. Thus an out of control condition on the means chart should then not be considered as a shift in the sample mean. In this sense the range chart will generally require the first attention. If no special causes are present, or special causes have been identified, the chart for the means should be (re-)examined.

8·3.2 *Situation II*

Extra variation between samples typically arises in batch processes, but may also occur in other situations. Instead of the model (8.7) the following holds for measurements X_{tj}:

$$X_{tj} = \mu + B_t + E_{tj} \quad \text{for } j = 1, 2, \dots, n \text{ and } t = 1, 2, 3, \dots , \tag{8.11}$$

where μ indicates the constant overall mean, B_t models the random variation between samples (batches) and E_{tj} models the random variation within samples. The E_{tj} are assumed to be mutually independent and normally distributed with means 0 and variance σ_e^2; the B_t are also assumed to be independent, normally distributed with means 0 and variance σ_b^2. Thus the standard deviation of sample means $\overline{X}_t = \sqrt{(\sigma_b^2 + \sigma_e^2/n)}$. The distribution of the R_t is not affected by the between sample variation; the control chart for the sample ranges can remain the same. A robust 'omnibus method' to establish limits for the sample means $\overline{X}_t$ is to treat these $\overline{X}_t$ as individual measurements and apply the methods of section 8.2. So let

$$MR_{\overline{X}} = \frac{1}{k-1}\sum_{t=2}^{k} |\overline{X}_t - \overline{X}_{t-1}|,$$

then the limits for the X-chart are (where $\overline{\overline{X}}$ is the average of all sample means):

$$\begin{aligned} \text{UCL}_{\overline{X}} &= \overline{\overline{X}} + 2.66MR_{\overline{X}}, \\ \text{CL}_{\overline{X}} &= \overline{\overline{X}}, \\ \text{LCL}_{\overline{X}} &= \overline{\overline{X}} - 2.66MR_{\overline{X}}. \end{aligned} \tag{8.12}$$

In the example of the car door silencers $MR_{\overline{X}} = 0.042$. Hence $(\text{LCL}, \text{UCL}) = (9.395, 9.619)$, which does not differ significantly from the limits established under model (8.7). This confirms that it is safe to assume that no extra between sample variation is present. A rough test for the presence of extra between sample variation (based on an approximate F-test to compare variances) can be derived from $MR_{\overline{X}}$ and $\overline{R}$:

> *If* $2.66MR_{\overline{X}}/\overline{R} > 1.225$ *then extra between sample variation should be taken into account.*

This simple rule might be used to decide upon the control chart limits calculation method to be used.

8·4 Control Charts for the Standard Situation

The examples of grinding wafers and glass deposition for diodes discussed in Chapter 7 clearly show the features of what we called situation IV: systematic effects within a sample inherent to the process (common cause) and extra between sample variation. For the wafer grinding example this was established with a multi-vari chart (Figure 7.7) and more formally by analysis of variance (Table 7.3). This suggests the following general model for an under control process, in which fixed differences within a sample as well as random between batch variation are to be considered as common cause variation:

$$X_{tj} = \mu + w_j + B_t + E_{tj}\,, \tag{8.13}$$

where X_{tj} is the measurement of sample t (*e.g.*, batch) of within sample product j (*e.g.*, the wafer in a batch). On the right hand side $\mu + w_j$ is the fixed mean level of product j ($\sum_{j=1}^{n} w_j = 0$), B_t models the random between batch variation, and finally E_{tj} models the remaining random variation. It is assumed that the B_t and E_{tj} are mutually stochastically independent variables, normally distributed with means 0 and variances σ_b^2 and σ_e^2, respectively. In Roes and Does (1995) control charts based on this model and example are thoroughly discussed; Does, Roes and Trip (1998) favourably place the resulting method in a broader multivariate context.

In many instances in which model (8.13) is appropriate, monitoring of the process can be focused on the batch mean, and certain differences within a batch. These differences need to be linked with well known characteristics of the process, in order to understand out of control situations and remove or correct the corresponding special cause of variation.

The suggested approach implies that the sample mean μ, and specific *linear contrasts* γ of the fixed effects w_j in model (8.13) are to be controlled:

$$\gamma = \sum_{j=1}^{n} c_j w_j, \quad \text{with} \quad \sum_{j=1}^{n} c_j = 0.$$

If there is more than one linear contrast, the contrasts are preferably chosen *orthogonally*:

$$\text{If} \quad \delta = \sum_{j=1}^{n} d_j w_j \quad \text{and} \quad \epsilon = \sum_{j=1}^{n} e_j w_j \quad \text{then} \quad \sum_{j=1}^{n} d_j e_j = 0.$$

Then their sample estimates are both mutually independent and independent from the sample mean, which greatly simplifies interpretation.

8·4.1 Control Chart for the Mean

The sample means are denoted by $\overline{X}_t = (1/n)\sum_{j=1}^{n} X_{tj}$. The mean and variance of $\overline{X}_t$ are given by $E(\overline{X}_t) = \mu$ and $\text{var}(\overline{X}_t) = \sigma_e^2/n + \sigma_b^2$. This variance is optimally (*i.e.*, , minimum variance unbiased) estimated by the estimator based on the analysis of variance:

$$S_b^2 = \frac{1}{k-1}\sum_{t=1}^{k}(\overline{X}_t - \overline{\overline{X}})^2. \tag{8.14}$$

From (8.14) it is clear that the batch means can be treated as individual observations. As is well known, S_b^2 is very sensitive to possible trends or jumps in the batch averages. With a relative efficiency of approximately 61% (section 8.2), the mean of the moving ranges $MR_{\overline{X}}$ provides a good alternative:

$$MR_{\overline{X}} = \frac{1}{k-1}\sum_{t=2}^{k}|\overline{X}_t - \overline{X}_{t-1}|.$$

As before, $MR_{\overline{X}}/d_2$ is an unbiased estimator of $(\sigma_e^2/n + \sigma_b^2)^{1/2}$. Control limits for the mean are given by (8.15). In the case S_b/c_4 is used, this has an obvious analogue.

$$\begin{aligned} \mathrm{UCL}_{\overline{X}} &= \overline{\overline{X}} + 3\frac{MR_{\overline{X}}}{d_2}, \\ \mathrm{CL} &= \overline{\overline{X}}, \\ \mathrm{LCL}_{\overline{X}} &= \overline{\overline{X}} - 3\frac{MR_{\overline{X}}}{d_2}. \end{aligned} \tag{8.15}$$

8·4.2 Control Charts for Linear Contrasts

The linear contrast $\gamma = \sum_{j=1}^{n} c_j w_j$ is estimated in the t-th sample by $C_t = \sum_{j=1}^{n} c_j X_{tj}$.

The mean and variance of C_t are given by:

$$\begin{aligned} E(C_t) &= \gamma, \\ \mathrm{var}(C_t) &= \left[\sum_{j=1}^{n} c_j\right]^2 \sigma_b^2 + \left(\sum_{j=1}^{n} c_j^2\right)\sigma_e^2. \end{aligned} \tag{8.16}$$

Since $\sum_{j=1}^{n} c_j = 0$, $\mathrm{var}(C_t) = (\sum_{j=1}^{n} c_j^2)\sigma_e^2$, and this does not depend on the between batch variation, which is also intuitively clear. Hence the optimal unbiased estimator of the standard deviation of C_t is $(\sum_{j=1}^{n} c_j^2)^{1/2} S_e/c_4$, with S_e given by:

$$S_e^2 = \frac{1}{(n-1)(k-1)}\sum_{j=1}^{n}\sum_{t=1}^{k}(X_{tj} - \overline{X}_{t.} - \overline{X}_{.j} + \overline{X}_{..})^2, \tag{8.17}$$

and $c4 = c_4(df+1)$ (with $df = (n-1)(k-1)$) is the constant to make the estimator unbiased (cf. Duncan (1986)). S_e^2 is the mean square error from the analysis of variance of model (8.13). In contrast to S_b^2, S_e^2 is not sensitive to jumps or trends in the batch averages, provided they are approximately equal for all wafers. Thus

S_e is a very sensible estimator to use in many situations. It follows that normal theory optimal limits for linear contrasts are given by (8.18).

$$\begin{aligned} \mathrm{UCL}_\gamma &= \overline{C} + 3\sqrt{\sum_{j=1}^{n} c_j^2}\,\frac{S_e}{c_4}, \\ \mathrm{CL}_\gamma &= \overline{C}, \\ \mathrm{LCL}_\gamma &= \overline{C} - 3\sqrt{\sum_{j=1}^{n} c_j^2}\,\frac{S_e}{c_4}. \end{aligned} \tag{8.18}$$

There can be practical and statistical reasons to use moving range based estimators, as with the batch mean. Let C_t denote a sample contrast as before and define:

$$MR_C = \frac{1}{k-1}\sum_{t=2}^{k} |C_t - C_{t-1}|.$$

It is easily verified that MR_C/d_2 is an unbiased estimator of $(\sum_{j=1}^n c_j^2)^{1/2}\sigma_e$ and control limits for C_t can be determined exactly as for the batch mean, defined in (8.15). The variance of MR_c/d_2 is approximately $(\sum_{j=1}^n c_j^2)(0.8264k - 1.082)/(k-1)^2$ (Cryer and Ryan (1990)), which when compared with the variance of $(\sum_{j=1}^n c_j^2)^{1/2} S_e/c_4$ leads to a relative efficiency of approximately $0.61/(n-1)$.

Example (continued): Grinding Silicon Wafers

The grinding wafers example introduced in Chapter 7 was one of the triggers for developing control charts for model (8.13). To illustrate the procedures as described control chart limits are calculated from the data in Table 7.2. The basic assumptions of model (8.13) were checked by means of a normal probability plot of the residuals and by comparing the residual variances between positions and between batches. No violations of the assumptions of normality and equal variances could be detected. The parameters of interest are:

(a) The mean μ, to control the average thicknsss of a batch;

(b) The difference between inner and outer circle $\delta = -\frac{1}{2}w_1 - \frac{1}{2}w_2 + w_5$;

(c) The difference on the middle circle $\epsilon = w_3 - w_4$.

It is easily verified that δ and ϵ are indeed orthogonal contrasts. Both contrasts δ and ϵ are the result of a principal component analysis of two weeks measurement of full batches, and physical explanations for these components.

Control Chart for the Mean

The batch mean $\overline{X}_t$ measures whether or not the average thickness is constant. Out of control conditions on the mean can be resolved by adjustment through a *control knob*, regulating the resulting average thickness. Control limits for the $\overline{X}_t$ calculated using (8.15) are $(\mathrm{LCL}_{\overline{X}}, \mathrm{CL}_{\overline{X}}, \mathrm{UCL}_{\overline{X}}) = (238, 245, 252)$.

Control Charts for the Contrasts

The two weeks period indicated a large and persistent difference between the inner and outer circles. The reason for this is a small but unavoidable displacement of the grindstone relative to the grinding table and small differences in horizontal positions of the wafers. Out of control conditions may be caused by a change in this displacement, but more likely by the accumulation of dirt which, as a result of grinding many batches, deposits on the grindstone and/or the plane where the wafers are positioned. The control chart for the contrast δ is a graph of the statistic $D_t = -\frac{1}{2}X_{t_1} - \frac{1}{2}X_{t_2} + X_{t_5}$ versus t, with limits (based on the mean moving range): $(\text{LCL}_D, \text{CL}_D, \text{UCL}_D) = (-1.2, 5.4, 12.0)$.

The final important fixed effect contrast turned out to represent the differences within the middle circle, although these differences contributed only a relative small part of the total variation. Out of control conditions on this difference are harder to pinpoint, but they do lead quite often to different technical adjustments of the grinder (such as adjustments of spin speed, or oil pressure). The control chart limits for the statistics $E_t = X_{t_3} - X_{t_4}$ which monitor this contrast ϵ are $(\text{LCL}_E, \text{CL}_E, \text{UCL}_E) = (-10.6, -3.0, 4.6)$.

The control charts for this example are shown in Figure 8.3. The chart for the mean indicates a stable mean thickness. An out of control situation occurs when there is a difference between the inner and outer circle in the final batches (if supplementary run rules are applied). It turned out that grindings kept sticking to the grindstone at this point causing the increase in this difference. Dressing or even, if necessary, replacing the grindstone is typically needed to resolve the special cause.

Batches 13 and 23, and batch 8 nearly, went out of control on the difference within the middle circle. The wafer from position 18 was thicker than the wafer from position 19, which usually is not the case. This might well be an accidental interchange of the two wafers—which happens easily. But it can also point to more serious problems, such as changed horizontal positions of the wafers caused by dirt. Then a cleaning of the system is called for.

Example (continued): Medium Power Rectifiers

The second case in this section also originates from Philips Stadskanaal in the Netherlands, and continues the glass deposition process for medium power rectifiers (Chapter 7). Recall that the products are placed in a holder according to Figure 7.10 during and after the glass deposition process. Each shift one holder is sampled, and the diameters of 7 products are measured. After five weeks of data collection 95 samples were available for analysis. The correlation matrix of the diameters clearly shows within sample relationships (see Table 8.7).

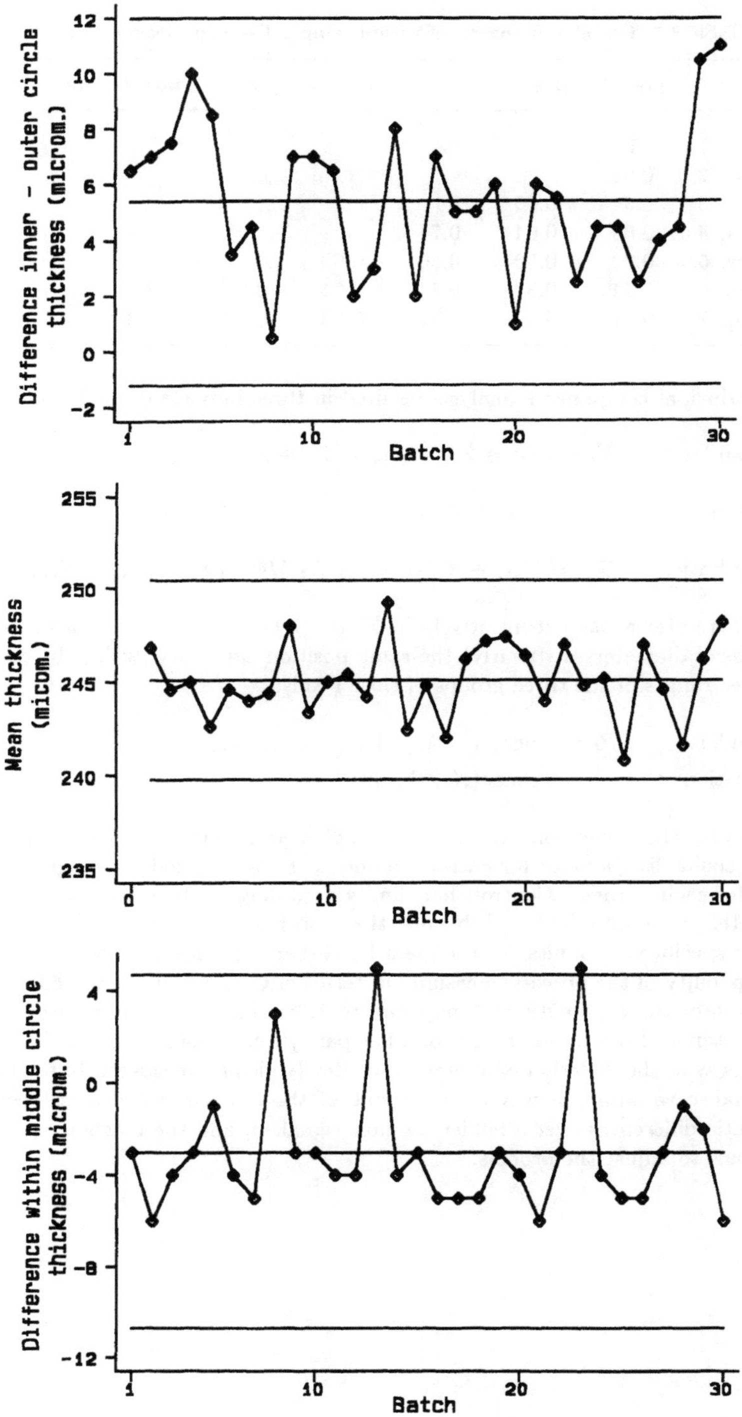

Figure 8.3: Control charts for grinding silicon wafers.

Table 8.7: Correlation matrix of within sample diameter measurements.

	pos. 1	pos. 2	pos. 3	pos. 4	pos. 5	pos. 6	pos. 7
pos. 1	1						
pos. 2	0.65	1					
pos. 3	0.34	0.60	1				
pos. 4	0.39	0.64	0.74	1			
pos. 5	0.32	0.50	0.50	0.63	1		
pos. 6	0.24	0.33	0.54	0.64	0.57	1	
pos. 7	0.21	0.32	0.51	0.53	0.40	0.73	1

A principal components analysis resulted in three important components:

(a) Mean level: $\overline{X}_t = (X_{t_1} + X_{t_2} + X_{t_3} + X_{t_4} + X_{t_5} + X_{t_6} + X_{t_7})/7$;

(b) Linear: $D_{\mathrm{L}t} = (X_{t_1} + X_{t_2})/2 - (X_{t_6} + X_{t_7})/2$;

(c) Quadratic: $D_{\mathrm{Q}t} = (X_{t_1} + X_{t_2} + X_{t_6} + X_{t_7})/4 - (X_{t_3} + X_{t_4} + X_{t_5})/3$,

which were also explained intuitively in Figure 7.11. The random variation between bead diameters at (nearly) the same position can be controlled by adding the three ranges of the three groups of close beads:

(d) Random variation: $R_t = \mathrm{range}(X_{t_1}, X_{t_2}) + \mathrm{range}(X_{t_3}, X_{t_4}, X_{t_5}) + \mathrm{range}(X_{t_6}, X_{t_7})$.

Because these components could be explained in physical terms, and the OCAP could discriminate for each component, it was decided to design control charts for each of them. Control chart limits according to the methods presented in chapter are summarised in Table 8.8; the charts are displayed in Figure 8.4.

The specification limits for the glass bead diameters are 3.76 and 4.71 mm. The capability of the process, measured in terms of C_{pk}, initially was 1.53. A few months later the capability had improved to 1.68. The improvement owed to a smaller standard deviation (Figure 8.5) has partly been offset by a shift from the target away in the direction of a larger diameter (which is considered to be safer). The smaller variation, however, is a result of the new control charts, because systematic differences over a holder are now signalled, and the machine operator has means to adjust the process.

Table 8.8: Control chart limits for glass deposition medium power rectifiers.

Component	Estimate from initial sample ($n = 7$, $k = 95$)		Control chart limits		
	Mean	*MR*	LCL	CL	UCL
$\overline{X}_t$ (mean)	4.236	0.059	4.077	4.236	4.394
R_t (random variation)	0.205	0.111	0	0.205	0.501
D_t (linear)	0.001	0.135	−0.358	0.001	0.360
D_{Qt} (quadratic)	−0.026	0.052	−0.162	−0.026	0.111

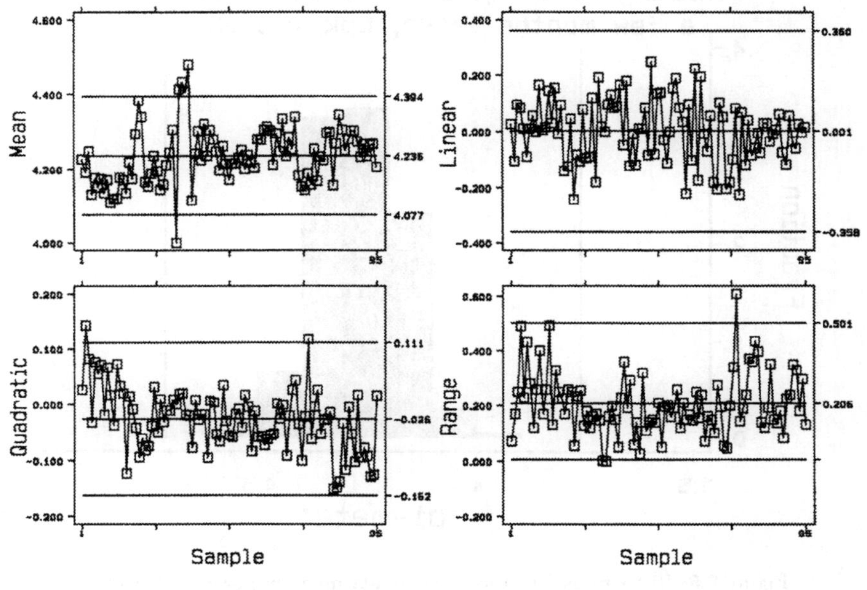

Figure 8.4: Control charts for glass deposition medium power rectifiers.

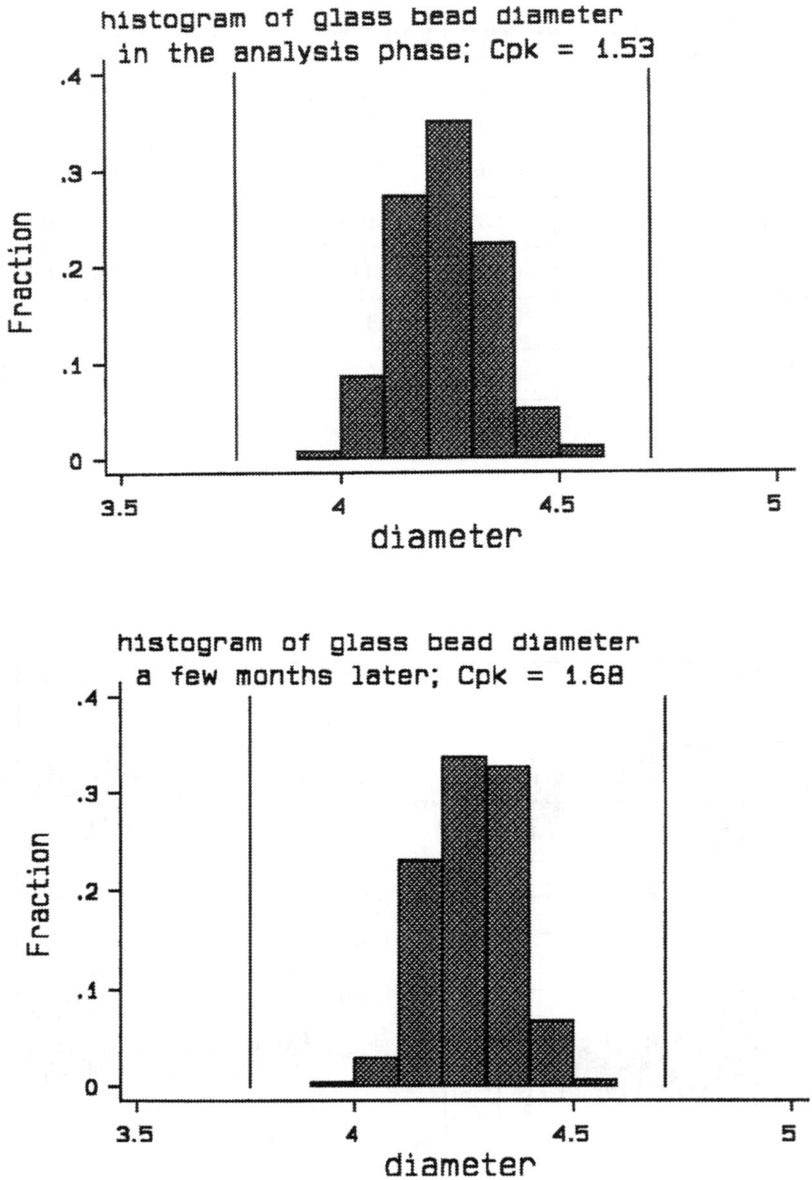

Figure 8.5: Histograms for process capability medium power rectifiers.

8·5 Control Charts in Low Volume Manufacturing

8·5.1 *Introduction*

Today's manufacturing environment hardly resembles the high volume production of the twenties, in which period Statistical Process Control methods were introduced. An enormous flexibility in production processes is now required because of the diversity of products, shorter product life cycles, and products tailored to specific customer requirements. Current manufacturing practice is typically characterized either by frequent set up changes to accommodate a wide range of different (but basically similar) products, or by intrinsically low production volumes of the same type of product. Printed circuit boards for specialized applications (*e.g.*, weapon systems and aircraft) are an example of the first situation, the manufacturing of wafer steppers for semiconductor production is an example of the latter. Thus such production processes can be characterized as either ***short run*** processes which produce a high volume of similar products, but which require frequent changes in set up (*e.g.*, temperature, pressure, process time), or ***low volume*** processes which produce a volume of the same products that accumulates slow over time (roughly one per day or less) and for which the degree of repetition of any process step is small. A key issue in low volume processes is that accumulation of control data is slow, and environmental conditions, operating procedures, and other relevant factors are more likely to change from one observation to the next. In both cases control charting methods require adaptations to be effective. Two types of adaptation have extensively been dealt with by several authors in recent years. These are ***standardization*** and methods to obtain ***exact control chart limits*** which retain the pre-specified probability of a point falling outside the control limits when these limits are estimated.

The standardization methods commonly used are variations of two basic approaches:

- The *difference from nominal* approach. A specific nominal value of the process (*e.g.*, a target thickness) is subtracted from each measured value. The differences thus obtained can be charted in the order of processing, provided that process variability is the same across the different process settings.

- The ***stabilization*** approach. To resolve non-constant process variability across different process settings, differences from nominal are standardized with the corresponding standard deviations.

The nominal values used can also be estimated from historical stable mean levels, which is, in fact, to be preferred if it is difficult to shift the process exactly on target. Most of the standardization methods are described in Wheeler (1991). Alternative forms of standardization are discussed in Al-Salti, Aspinwall and Statham (1992), and Farnum (1992) includes measurement error in different models to arrive at the proper form of standardization. These methods have proved to be most useful in short run processes as defined above. Since this approach will usually require availability of sufficient data to estimate historic mean levels and process variability, it is not applicable to the true low volume case. Also, the charts discussed in the previous section require a sufficiently large initial sample to be applied in manufacturing processes (see also Chapter 6).

This section deals with the true low volume case. The Q charts developed by Quesenberry (1991a) for variables and for Binomial, Poisson, and Geometric parameters in Quesenberry (1991b, 1991c and 1995a) provide an elegant method to establish both standardization and exact limits. The Q chart for variables is introduced, in the case of unknown parameters and individual observations—the most important application in our view. Tests for special causes are considered, as well as alternative options of charting an exponentially weighted moving average of Q statistics and tightening control limits. Q statistics based on the mean moving range as estimator of the process standard deviation result in the Q chart.

8·5.2 The Q Chart

As stated, individual measurements with unknown mean and standard deviation are considered most important for practical applications. To summarize the Q chart for this case, let $X_1, X_2, X_3, \ldots$ denote the sequence in time of measurements of the process or product characteristic used for monitoring a low volume process. For an in control process, the X_t distributed as $N(\mu, \sigma^2)$ for all t, with μ and σ unknown. Now, let:

$$\begin{aligned} \overline{X}_t &= \frac{1}{t}\sum_{i=1}^{t} X_i \\ &= \left(1-\frac{1}{t}\right)\overline{X}_{t-1} + \frac{1}{t}X_t, \end{aligned} \tag{8.19}$$

and (for $t > 2$)

$$\begin{aligned} S_t^2 &= \frac{1}{t-1}\sum_{i=1}^{t}(X_i - \overline{X}_t)^2 \\ &= \frac{t-2}{t-1}S_{t-1}^2 + \frac{1}{t}(X_t - \overline{X}_{t-1})^2. \end{aligned} \tag{8.20}$$

The statistics T_t given by

$$T_t = \sqrt{\frac{t-1}{t}}\,\frac{(X_t - \overline{X}_{t-1})}{S_{t-1}} \qquad (t > 2) \tag{8.21}$$

are known to be distributed according to Student's t, with $t-2$ degrees of freedom. Quesenberry (1991a) shows that the T_t are stochastically independent when the process is under control. Let the cumulative distribution of Students t with ν degrees of freedom be denoted by G_ν. Using the probability integral transform, it directly follows that $G_{t-2}(T_t)$ is distributed uniformly over the interval (0,1). Denoting the quantile function of the standard normal distribution by Φ^{-1}, it follows that:

$$Q_t = \Phi^{-1}(G_{t-2}(T_t)) \quad (t > 2) \tag{8.22}$$

is distributed according to the standard normal distribution $N(0, 1)$.

The statistics Q_t are the quantities to be charted, which by construction results in four important attractive properties:

- ☐ Fixed control limits at −3, 0 and +3 can be used starting at the third observation, and these limits ensure an exact false alarm rate of 0.0027 (two-sided).
- ☐ The Q_t statistics are independent and standard normal distributed. This ensures that the under control behaviour is stationary and can be well interpreted.
- ☐ The Q_t statistics provide the uniform most powerful unbiased test for detecting a shift at the next observation.
- ☐ The standardization can also be used to standardize chart statistics for different products or characteristics from the same process (short run applications).

Note that through applying the function $\Phi^{-1}(G_{k-2}(\cdot))$, the exact under control signal probability is achieved by smaller weights for the more extreme deviations from 0 of the T_t, instead of widening the control limits. Around 0 the function is close to linear and hence the resulting chart will show quite some similarity with charting the T_t directly.

In addition to the favourable characteristics, the Q chart suffers from drawbacks which are quite relevant for practical applications. The one generally considered most important concerns the fact that if a shift in the process mean is not detected quickly, the adaptive nature of the estimators of the process mean and variability causes the charted points to eventually settle into a pattern indicative of an under control process (Quesenberry, 1991a). Hence if an out of control condition with respect to a one step permanent shift in μ is not detected shortly after it occurred, it is unlikely it will be detected as such. Moreover, a linear trend may go totally undetected by the Q chart used with only the 1–1 test of one point falling outside the limits is used as signal, as was shown by Farnum (1995). Using arguments similar to his, this worrying result can even be extended to quadratic and cubic trends (Jonkers, 1997). The following charting alternatives for current control can improve performance in low volume situations:

- ☐ The Q chart with a suitable set of run rules chosen from the following four most powerful rules (see also Chapter 6):
  - ☐ 1–of–1 test (test 1): signals if the last point is beyond the control limits (±3);
  - ☐ 8–of–8 test (test 2): signals if eight consecutive points fall on the same side of the central line (0).
  - ☐ 2–of–3 test (test 5): signals if two out of the last three points are beyond the same warning limit (±2);
  - ☐ 4–of–5 test (test 6): signals if four of the last five points are beyond the same auxiliary limit (±1);
- ☐ Tightening control limits, which yields a higher false alarm rate but also a larger sensitivity to out of control situations. Limits could be set at ±2, ±2.5 and ±2.75 in addition to limits at ±3.

- An EWMA of the Q statistics. The EWMA statistic Z_t is defined for the third observation onwards as follows:

$$Z_t = \lambda Q_t + (1-\lambda)Z_{t-1}, \quad t = 3, 4, \ldots, \tag{8.23}$$

with $Z_2 = 0$. The Z_t are plotted against control limits at $\pm K\sqrt{\lambda/(2-\lambda)})$. The parameter λ set at 0.25 provides relatively good performance against a broad range of shift sizes. The value of K ($K = 2.5$) is chosen to obtain an under control average run length of 124 to match the under control behaviour with the combination of rules above, rather than with the 1–of–1 rule alone.

8·5.3 The QR Chart

The standard deviations S_t are well known to be heavily inflated when an outlier is present. Thus the influence of an out of control condition which is not detected at the next observation is particularly substantial for the adaptive Q charts. As before, the mean moving range as alternative estimator of the process standard deviation will reduce this influence. Hence once again let:

$$\begin{aligned} MR_t &= \frac{1}{t-1}\sum_{i=2}^{t}|X_i - X_{i-1}| \\ &= \frac{t-2}{t-1}MR_{t-1} + \frac{1}{t-1}|X - t - X_{t-1}|. \end{aligned} \tag{8.24}$$

Subsequently, the corresponding standardized statistics TR_t are given by:

$$TR_t = \sqrt{\frac{t-1}{t}}\,\tau_{t-1}\frac{(X_t - \overline{X}_{t-1})}{MR_{t-1}/d_2} \quad (t \geqslant 2). \tag{8.25}$$

The constants τ_t and ν_t are applied in order to arrange that $(MR_t/d_2)/\tau$ is approximately distributed as $\sigma\chi(\nu_t)/\nu_t^{1/2}$, where $\chi^2(m)$ denotes a χ-squared distribution with m degrees of freedom. The constants are given by (cf. Roes *et al.*1993):

$$\begin{aligned} \nu_t &= \frac{1}{2}\left(1 + \frac{1}{\text{var}(MR_t/\sigma d_2)}\right), \\ \tau_t &= \sqrt{1 + \text{var}(MR_t/\sigma d_2)}, \end{aligned} \tag{8.26}$$

The variance of MR_t/d_2 was derived by Cryer and Ryan (1990) to be about:

$$\text{var}\left(\frac{MR_t}{d_2}\right) = \frac{0.8264t - 1.082}{(t-1)^2}\sigma^2 \quad (t \geqslant 2). \tag{8.27}$$

As a consequence of the approximation above TR_t is approximately distributed as Student's t, with ν_{-1} degrees of freedom. Hence we can proceed as in (8.22) to find that the statistics QR_t:

$$QR_t = \Phi^{-1}(G_{\nu_{t-1}}(TR_t)) \quad (k \geqslant 2,\ \nu_{t-1} \geqslant 1) \tag{8.28}$$

are approximately standard normal distributed. All TR_t and TR_{t+s} for $s > 0$ are uncorrelated (Roes *et al.*1998), which thus holds approximately for QR_t and QR_{t+s}. Therefore QR charts with fixed limits can be used, starting at the observation t for which $\nu t - 1 > 1$. Loosely formulated, the QR statistics are thus approximately independent and standard normal distributed. The QR charts which can be constructed along these lines were indeed shown to be less sensitive to outliers (Roes, Does and Jonkers (1998)). Obviously, run rules and EWMA can be applied in the same way as for the standard Q charts.

Example (continued): Reticle Handling Elevators

ASM Lithography was already introduced as a real low volume manufacturer of wafer steppers. The process of mounting reticle handling elevators presents an application for QR charts. The 40 fall times from Table 7.1 were used to create the QR chart in Figure 8.6 (starting from the 10th elevator). A special cause is clearly indicated at the 33rd elevator. If extra run rules are used (specifically test 6: 4 out of 5 beyond the 1σ line), there is strong evidence that the average fall time has increased consistently beyond the 33rd elevator.

In this example, the standard Q chart looks virtually the same and leads to the same conclusions.

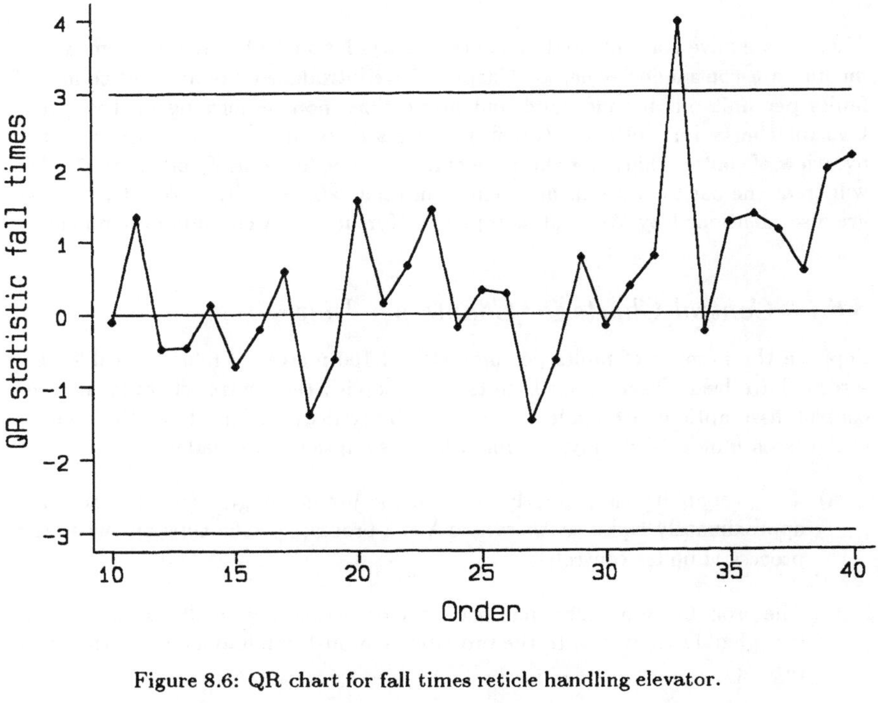

Figure 8.6: QR chart for fall times reticle handling elevator.

8·5.4 Guidelines

Guidelines for practitioners on the effective use of $Q(R)$ type charts in low volume processes are the following:

- □ The QR charts should be applied with a combination of tests for special causes. The combination of the 1–of–1, 2–of–3, 4–of–5 and 8–of–8 tests provide adequate performance.
- □ If the use on the shop floor can be made effective, an EWMA on the QR statistics provides the best overall performance.
- □ The use of the mean moving range will overall enhance performance, except at the first observation after the onset of the shift or trend.
- □ A trend chart should be shown along with the control chart of choice, to enhance immediate interpretation.

Performance results underlying this guideline for the alternatives investigated were obtained through simulation studies according to the same or higher standards as used by Quesenberry (1995b) and Del Castillo and Montgomery (1995).

8·6 Control Charts for Counts and Proportions

Thus far we have concentrated on control charts for variables, requiring measurements on a continuous scale. In Chapter 7 we introduced examples of counts of faults per unit on the one hand and proportions non-conforming on the other. Control charts for both are (briefly) addressed in this section. An extensive overview of control charts for this type of data can be found in Woodall (1997). We will treat the basics and indicate some issues not satisfactorily resolved yet; these are also mentioned by Woodall as topics for further research and development.

8·6.1 Control Charts for Counts per Unit

Suppose the number of faults per unit (PCB, 100 metres of fabric or cable, TV screen) have been observed on k units. To establish the control chart for current control, assumptions are made on the probability distribution of N_t, the number of faults on unit t. Typically, the following assumptions are made:

(a) The probability that exactly one fault occurs in a region (unit) of size h is approximately equal to λh, where λ is a process specific constant when the process is under control;

(b) The probability of more than one fault occurring in a small region of size h is negligible compared to the probability of just one happening in the same region;

(c) The number of faults in non-overlapping regions are independent.

It can be shown that under these assumptions the probability distribution of N_t (on a unit assumed to have size $h = 1$) is Poisson:

$$P(N_t = z) = e^{-\lambda z} \frac{(\lambda)^z}{z!}.$$

Based on this probability distribution exact control limits can be calculated. However, it is customary to do so based on the 3σ basis. The expectation and variance of a Poisson distributed variable both equal λ. Hence if the observed mean number of faults in the initial sample of size k is $\bar{c}$, the limits are:

$$\begin{aligned} \mathrm{UCL}_c &= \bar{c} + 3\sqrt{\bar{c}}, \\ \mathrm{CL}_c &= \bar{c}, \\ \mathrm{LCL}_e &= \bar{c} - 3\sqrt{\bar{c}}. \end{aligned} \tag{8.29}$$

In the case where the calculated LCL_c is negative the actual lower control limit will be set to 0. The resulting chart is known as the c-chart, where the c stands for 'counts'.

Usually the assumptions made are too restrictive to be realistic in manufacturing processes. This has already been indicated in Chapter 7: systematic effects within the unit as well as extra variation between units (over-dispersion) are rather rule and not an exception. In the case where the counts are not too small (on average larger than 5; all or most larger than 0) and the total sample size is sufficiently large (as indicated in Chapter 6), the mean moving range method of section 8.2 can be applied straightforwardly. For small counts generalized linear (mixed) modelling (Myers and Montgomery, 1997) may be helpful—there is not much practical experience to date.

Example: Mounting Components on Printed Circuit Boards

At final assembly of Printed Circuit Boards (PCBs) (part of the) components are soldered onto the boards. The number of components per 10,000 soldered that does not make electrical contact was used by the PAT as measure to monitor the mounting process. Data collected on 30 samples of 10,000 mounted components are displayed in Table 8.9.

Control chart limits can be calculated using (8.29) and alternatively by applying the mean moving range method. In this case both methods diverge considerably (Table 8.10), which is owing to extra common cause variation between the consecutive samples.

The (retrospective) control chart based on the mean moving range (Figure 8.7) signals one out of control condition at the 30th sample. The 'standard' limits would have signalled at least 3 more. If extra between sample variation is present the standard limits will lead to a higher false alarm rate.

8·6.2 Control Charts for Proportions

In samples of size n the proportion of non-conforming products (or sub-assemblies, parts) can be used to monitor a particular process. The p-chart for ***proportions*** could then be applicable. The assumptions underlying the standard p-chart are the following:

Table 8.9: Number of components not making electrical contact per 10,000 mounted.

t	c_t	t	c_t	t	c_t	t	c_t	t	c_t	t	c_t
1	6	6	3	11	12	16	32	21	11	26	9
2	8	7	7	12	2	17	7	22	7	27	29
3	8	8	4	13	30	18	1	23	14	28	11
4	4	9	2	14	16	19	1	24	18	29	8
5	20	10	2	15	16	20	8	25	8	30	38

Table 8.10: Control chart limits for PCB example.

Control limit	Standard (8.29)	Mean moving range method
UCL	22	36
CL	11	11
LCL	0	0

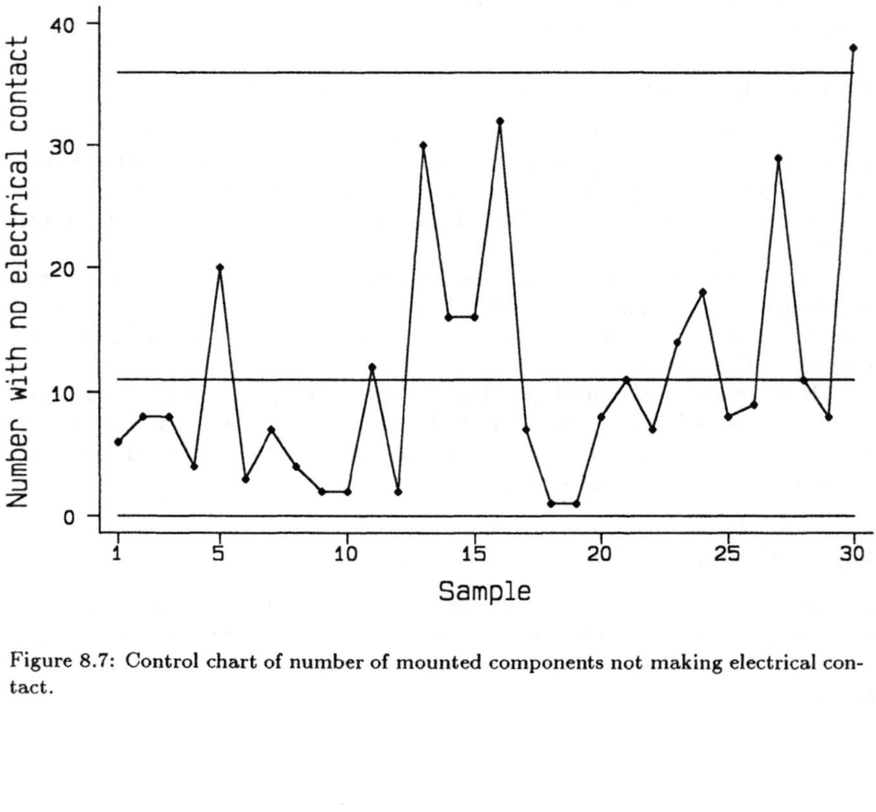

Figure 8.7: Control chart of number of mounted components not making electrical contact.

(a) Every single product has a constant probability p of becoming non-conforming, when the process is under control;

(b) Outcomes (conforming/non-conforming) for separate products are independent;

(c) Production volume is large compared to the samples drawn.

These assumption lead to a Binomial distribution for the number NC_t of non-conforming products in samples of size n:

$$P(NC_t = k) = \binom{n}{k} p^k (1-p)^{n-k}. \tag{8.30}$$

For samples sizes sufficiently large (25 and above), control limits are calculated from the observed mean proportion non-conforming $\bar{p}$ as follows:

$$\begin{aligned} \text{UCL}_p &= \bar{p} + 3\sqrt{\frac{\bar{p}(1-\bar{p})}{n}}, \\ \text{CL}_p &= \bar{p}, \\ \text{LCL}_p &= \bar{p} - 3\sqrt{\frac{\bar{p}(1-\bar{p})}{n}}. \end{aligned} \tag{8.31}$$

When the calculation gives a negative value the control limit is then set to 0 by hand. For the same reasons as with the c-chart, the assumptions are not always fulfilled. The mean moving range can again be applied, provided not too many samples with zero non-conforming are observed. Generalised linear (mixed) models can provide solutions for such cases.

8·7 Pre-Control

Pre-control is an alternative monitoring and control technique, originally developed by Satterthwaite (1954) and strongly promoted by Shainin (1984). As a follower of Shainin's ideas, Bhote (1991) considers control charts as 'horse and buggy' compared to the 'jet age' tool pre-control. Ledolter and Swersey (1997) present a more balanced view, which we follow in this section. The aim is to explain the basics and mechanics of pre-control and provide guidelines to practitioners upon the use of pre-control in relation to control charts.

In contrast to control charts, pre-control monitors the process against the specification interval. The following stepwise plan describes the mechanics of pre-control:

1 Pre-control divides the (two-sided) specification interval into four equal areas (Figure 8.8). Pre-control lines are drawn at the midpoints between the target and the lower (and upper) specification limit. The middle two areas (between the two pre-control lines) form the green zone. The two areas between pre-control lines and specification limits are the yellow zones. The areas outside the specification limits form the red zone.

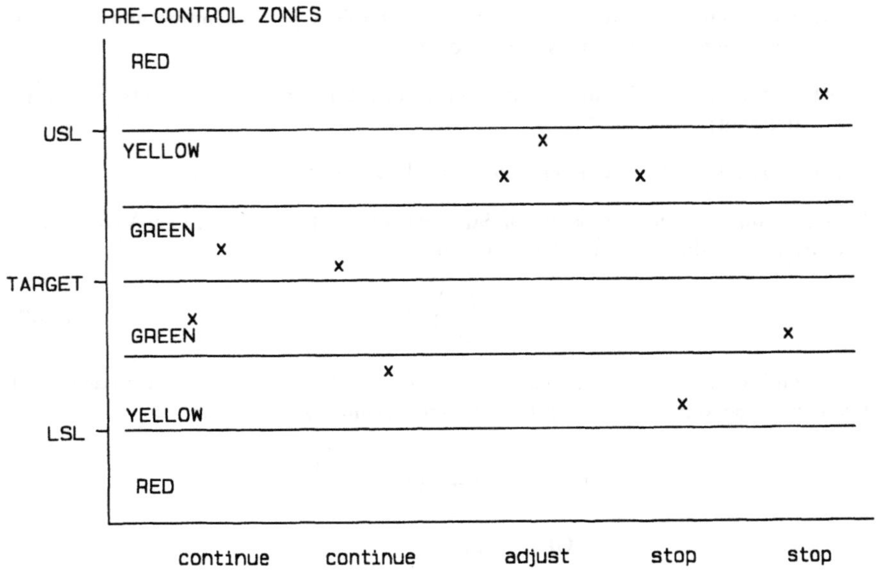

Figure 8.8: Pre-control zones and possible situations for two consecutive measurements.

2 Initial process capability is assessed by taking five consecutive products from the process. If the measurements of all five fall within the green zone, production can be started (or continued). In case one or more measurements fall outside the green zone, process variation needs to be reduced first.

3 During operation two consecutive products are sampled and measured at regular intervals. The decision rules for these samples are:

3a If both measurements fall in the green zone, production is continued.

3b If one measurement falls in the green zone and the other in the yellow zone, production is also continued.

3c If both measurements fall within the same yellow zone, the process is stopped and adjusted to target. If they fall in opposite yellow zones, 3the process is stopped and a thorough investigation is started to remove the cause of the excess variation.

3d If any measurement falls in the red zone the process is stopped and a thorough investigation is started to remove the cause of this defective.

At any point where the process is stopped (except for adjustment to target), process capability needs to be established according to step 2 before production can continue.

Pre-control is indeed simple. It is probably also effective if the process capability is adequate: well capable of producing within the specification limits (C_{pk}

at least 2). Otherwise, pre-control results in too many false alarms for which the process is stopped, investigations are started or the setting is tampered with. In line with the philosophy followed in this book, the excess variation responsible for the alarms in this case is common cause variation. As such, it is the responsibility of management to improve the system. Pre-control puts the burden on the operator, but the causes are beyond his control.

Another point to note is that for particular processes it may very well be economically viable to produce some defects (initially). In the case where pre-control is applied such processes are not even started!

The strong pressure for investigation and improvement evoked by applying pre-control is a very positive asset, that generally is included in the application of Statistical Process Control along the principles of this book as well.

Practitioners may benefit from pre-control is specific situations. It will not replace intelligent use of control charts (Ledolter and Swersey, 1997), but rather supplement them in well defined circumstances. These are defined by:

1 A high process capability ($C_{\rm pk}$ at least 2);

2 The measurements monitored with pre-control are causally linked to rejects;

3 The specifications on the measurements are well defined and unambiguous;

4 The sampling scheme of two consecutive items at regular intervals can sensibly implemented (which for most examples in this book is difficult).

These guidelines in fact clearly demonstrate the origin of pre-control: machining operations with straightforward monitoring and feedback. Extensions to more complex processes need further study. For the present we conclude that pre-control can be a useful supplement to the use of control charts in specific processes, without being an adequate substitute.

Chapter 9

Tools for Solving Problems

9·1 Introduction

This chapter discusses some tools which teams might use fruitfully when searching for improvements. Often a brainstorming session will give several clues for improvements, but occasionally the team will not get beyond conjectures. It then has to resort to other structured techniques.

One way to generate ideas is to analyse measurement data, log books, products, *etc.*. A careful examination might indicate relations, changes, differences, absurdities; in short, variation might be discovered and pointed at. Whether or not this will happen depends strongly on the used material. With simply available measurements, log books, products, *etc.*, chances are high that they will not give much information. For meaningful data it is nearly always necessary to collect your material in a well planned way.

The tools we discuss in this chapter constitute a collection of tools aimed at finding the culprit (the most important cause) and his possible accomplices. This collection is designed by Shainin, combining existing techniques with newly invented ones (Bhote, 1991). The system for finding 'the red X' (and 'the pink X' and 'pale pink X') is schematically shown in Figure 9.1.

The 'Components search', 'Multi-vari chart', and 'Pairwise Comparison' are aimed at quickly reducing a large number of possible causes to a smaller subselection comprising the culprit. 'Variables search' distinguishes important causes from less important ones, and the 'Factorial designs' show the combined effects of important variables (interactions). Finally, the assumed solution is checked against the current situation with 'B vs C'.

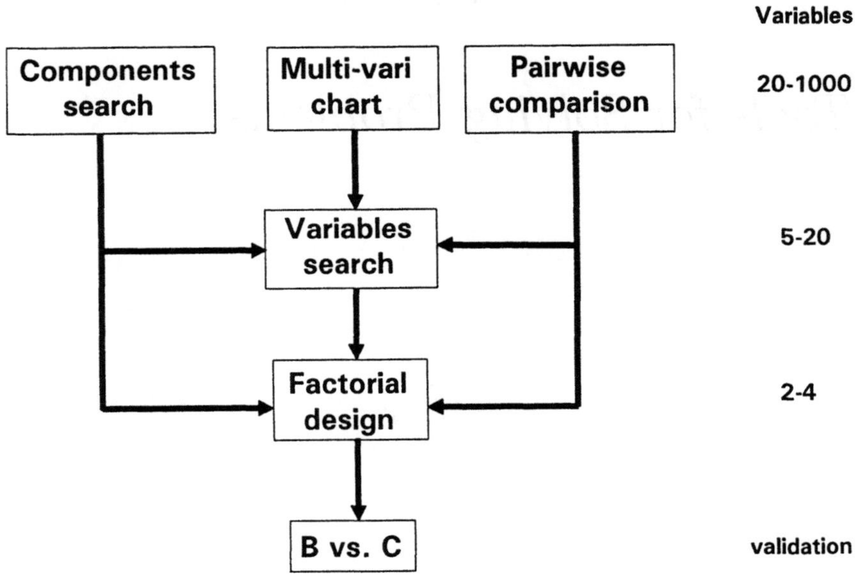

Figure 9.1: The system of problem solving tools.

9·2 Multi-Vari Chart

The most important source of variation is determined by a graph of a limited number of sensibly collected measurements. Distinction is made between three main sources of variation:

- Positional (within a single product, within a furnace batch, ...);
- Cyclical (between successive products, from batch to batch, ...);
- Time (from hour to hour, shift to shift, day to day, ...).

The measurements are collected in a stratified sample. Several measurements are taken from 3–5 successive products. Some time later (*e.g.*, one hour) the procedure is repeated, and after another period again. One may stop when the nature of the variation becomes clear.

Case Illustration: Tin Plating of Diodes

To solder electrical components onto a PCB the components need a certain amount of tin. Surface mounted diodes are therefore provided with a small tin layer on the flanges. The variation in the thickness is examined by sampling five products from each of three consecutive batches. The thickness of the layer is measured at three different positions at the flange, two at the edge, one in the middle. The results (in μm) are shown in the multi-vari chart of Figure 9.2. The chart shows that the variation between products, in the same batch, is the most important source, followed by the variation between batches. The trained observer will notice that the second measurement (the one in the middle) is nearly always 1 μm smaller that the two measurements near the edge. The conclusion is that for reducing the variation in thickness, differences between products within a single batch should be reduced first.

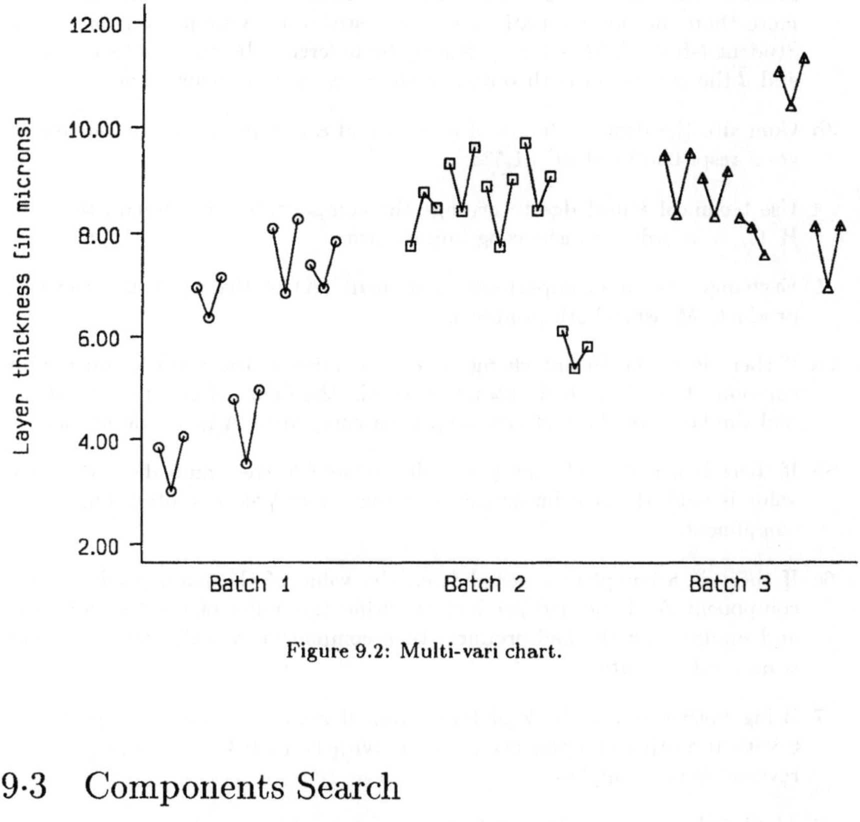

Figure 9.2: Multi-vari chart.

9·3 Components Search

In an assembly process the most important components can be traced with a 'components search'. To use this tool it is necessary that:

- □ the output of the product is measurable;
- □ products can be disassembled and assembled without significant change of output;
- □ at least one 'good' and one 'bad' product are available.

The idea behind components search is that interchanging components will lead us to the most important ones.

The search consists of ten steps:

1 Select the best and the worst product (from one day's production, for example).

2 Measure both products.

3a Disassemble and assemble the good product, measure it, and repeat this process. Do the same with the bad product. Continuation is only useful when all measurements of the good product are better than all measurements of the bad product. The difference between good and bad should be more than the noise variation; this is tested with a simple version of the Student t-test: $D/\overline{d} > 1.25$ (with D the difference between both medians, and $\overline{d}$ the average of both ranges of three repeated measurements.

3b Compute the decision limits of a good and a bad product: the median of good resp. bad product $\pm 1.53\overline{d}$.

4 Use technical knowledge to arrange the components; denominate them A, B, C, ... in order of decreasing importance.

5 Exchange the most important component (A) of the good and the bad product. Measure both products.

6a If there is no significant change (*i.e.*, the value of the good product with component A of the bad product is within the limits of the good product, and similarly for the bad product), then component A is not important.

6b If there is a partial change (one value is outside the limits, but the other value is not), the A is important, but there is at least one other important component.

6c If there is a complete reversal (*i.e.*, the value of the good product with component A of the bad product is within the limits of the bad product, and similarly for the bad product) then component A is the culprit. There is no need to continue.

7 Bring both products back in their original states and repeat steps 5 and 6 with the other components (starting with B, then C, ...) as long as the reversal is not complete.

8 The family of the most important components has been found.

9 Exchange the family of most important components of the good and the bad product; the reversal should then be complete.

10 The effects of the components are quantified by means of a factorial analysis (see the case illustration).

Case Illustration: Improving Measurement Equipment

The components search is especially meant for examination of electrical circuits (and other assembled products). The case illustration, however, is an illustration of another application of the method: the comparison of two 'identical' machines.

For measuring electrical characteristics of diodes several parallel machines are used, all of the same supplier, but of different age and therefore all somewhat different. Yet the yield of one machine is generally smaller than the yields of the other machines, even though the batches of products are randomly distributed over the machines. All kinds of maintenance did not improve this machine, and finally a team was given the assignment to solve this matter. They decided to attack the problem systematically, with a components search. The components of the bad machine were exchanged with the components of a good machine (of about the same age) in search of the culprit component. All possible components were listed and arranged from important to less important. The list comprised several transport magnets, the magnet bin, the measurement cage, and some transport wheels. They are denominated A, B, C, ... with A the most important component.

A batch of products was measured on both machines, and the reject percentages were determined. Then both machines were disassembled and assembled, and a new batch of products was measured. This was repeated and the reject rates were:

'Bad' Machine		'Good' Machine	
3.3%		1.4%	
3.5%	Median 3.3%	1.8%	Median 1.5%
2.5%	Range 1.0%	1.5%	Range 0.4%

All reject rates of the 'good' machine are lower than those of the 'bad' machine; the difference of the medians (D) is 1.8% and the average range ($\overline{d}$) is 0.7 %. The ratio $D/\overline{d} = 1.8/0.7$ ($= 2.6$) is clearly indicating a significant difference between both machines. The decision limits are:

'bad' machine: $3.3\% \pm 1.53 \times 0.7\%$ (*i.e.*, 2.23% and 4.37%)

'good' machine: $1.5\% \pm 1.53 \times 0.7\%$ (*i.e.*, 0.43% and 2.57%)

Then the components of both machines were exchanged, beginning with component A. The notation $A_B R_G$ indicates the combination of component A from the bad machine and the rest of the components from the good machine. The percentages rejects are (those outside the decision limits are printed bold):

$A_B R_G$: 2.3 %	$A_G R_B$: **2.1 %**	A has influence
$B_B R_G$: 1.3 %	$B_G R_B$: 4.0 %	B has no influence
$C_B R_G$: 1.1 %	$C_G R_B$: 3.2 %	C has no influence
$D_S R_G$: 1.9 %	$D_G R_S$: 2.7 %	D has no influence
$E_S R_G$: **3.4 %**	$E_G R_S$: **0.9 %**	E is important

Exchanging component E resulted in a complete reversal, so the search was stopped. To confirm the results, a final test was done in which both A and E were exchanged:

$A_B E_B R_G$: **3.6 %** $A_G E_G R_B$: **1.7 %** R has no influence

Conclusion: components A and E are important for the yield of the measuring machine, the influence of all other components is only limited. In Figure 9.3 the data of this experiment are summarized.

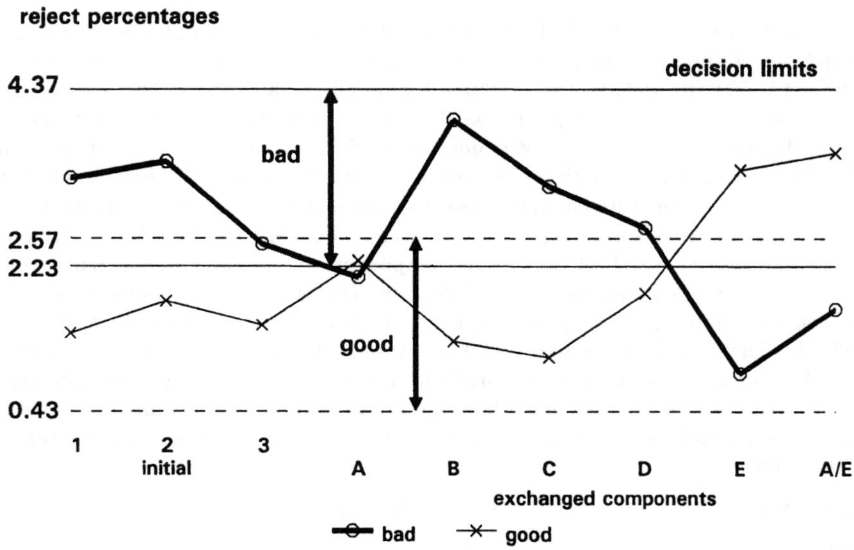

Figure 9.3: Components search of the measuring machines.

Finally the results are summarized in a factorial analysis, a table showing the effects of the important components, A and E in this case. All measurements are assigned to one of four possibilities $A_B E_B$, $A_B E_G$, $A_G E_B$, $A_G E_G$, determined by which components A and E were used. For each cell the median is calculated (see Table 9.1).

Table 9.1: Factorial analysis of the measuring machines experiment.

	A_B	A_G
E_B	3.3% 3.5% 2.5% 4.0% 2.7% 3.2% 3.6% Median = 3.3%	2.1% 3.4% Median = 2.75%
E_G	2.3% 0.9% Median = 1.6%	1.4% 1.8% 1.5% 1.3% 1.9% 1.1% 1.7% Median = 1.5%

The factorial analysis shows that component E has a large influence on the yield: using the good component nearly halves the reject percentage. The influence of component A is smaller. The interaction of both components is displayed in Figure 9.4. It shows that component A influences the yield only if the bad component E is used, with the good component E there is hardly any effect of component A.

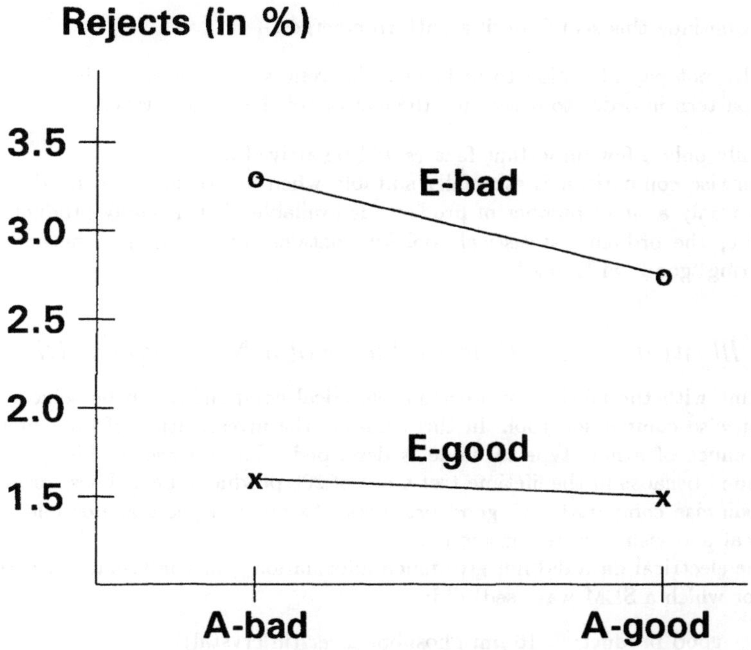

Figure 9.4: Interaction graph of components A and E.

Summary: This components search showed that component E is solely responsible for the lower yields of the 'bad' measuring machine. Further investigation of this part, one of the many transport magnets, showed its difference with the same parts in the other machines. A better control of the adjustment of this magnet assured the improvement of the yields.

9·4 Pairwise Comparison

If assembling and disassembling without effect on the output is not feasible, then the pairwise comparison is a substitute for components search. Again distinction should be made between 'good' and 'bad' (products, machines, ...), and for pairwise comparison a few examples of both kinds should be available. The idea behind this tool is that a pattern will emerge from a close examination of the differences between 'good' and 'bad'.

The procedure is a five step method:

1 Take a good and a bad product (made in the same period).

2 Examine this pair of products closely for differences (visual, dimensional, electrical, mechanical, chemical, ...). Use the eye, X-ray, SEM (Scanning Electron Microscope), destructive tests, *etc.*. Record the results.

3 Take a second pair of a good and a bad product, and repeat step 2.

4 Continue this search until a pattern emerges.

5 Do not pay attention to incidental differences, but focus on the repeating pattern in order to determine the sources of the biggest variation.

Generally only a few important factors will be arrived at.

Pairwise comparison is especially suitable when measuring is really difficult, or when only a small number of products is available. When many products are available, the ordinary statistical tool kit contains much stronger methods for comparing 'good' with 'bad'.

Case Illustration: Lifetime Problems of a New Type of Diode

Problems with the lifetime of a certain electrical component can be solved with the pairwise comparison tool. In this example the investigation of poor lifetime performance of a new type of diode is described. The release of this type was suspended because in the lifetime test 4 out of 200 products died. These products were pairwise compared with good products. As many aspects as possible, both electrical and visual, were considered.

The electrical data did not give much information, but the visual characteristics (for which a SEM was used) did:

pair 1	good product:	16 μm phosphor layer (in crystal)
	bad product:	10 μm phosphor layer
		dust particles in oxide layer
pair 2	good product:	19 μm phosphor layer
	bad product:	spots without phosphor layer
pair 3	good product:	21 μm phosphor layer
	bad product:	12 μm phosphor layer
		air inclusions in glass
pair 4	good product:	18 μm phosphor layer
	bad product:	8 μm phosphor layer
		dust particles in aluminium layer

The conclusion was that the thickness of the phosphor layer was the culprit: in bad products the layer was much thinner than in good products. Together with other contaminations this resulted in lifetime failures. The solution was to diffuse more phosphor in the crystal. This turned out to be the solution for the lifetime problems, and other parameters were not adversely affected at the same time.

9·5 Variables Search

Variables search is pre-eminently suited for separating important from unimportant variables. The method resembles components search (section 9.3), but is more powerful because it is not just limited to assembly processes.

The method consists of four phases:

I Selection.

II Elimination.

III Confirmation.

IV Analysis.

Phase I: Selection

1 Sort the mort important factors in descending order of perceived importance; name them A, B, C, ...

2 Designate a low level (L) and high level (H) to each variable, such that the best results are expected at the high level.

3 Run three experiments with all factors at their low level, and three with all factors at their high level. Let the order of the six experiments be random, because undesirable effects of unknown variables should be avoided as much as possible.

4 Check the following conditions:

- All high level results are better than the low level results;
- Calculate the median and range of the three high level results, and also of the low level results; then $D/\overline{d} \geqslant 1.25$ (with D the distance between the two medians and $\overline{d}$ the average of the two ranges).

The two sets of results are significant different if both conditions are fulfilled; then continue with step 7, otherwise go to step 5.

5 Exchange the low level and high level of the most important factor (A) and repeat step 3. Step 5 has to be repeated with the other factors as long as the difference between the high level results and the low level results is not significant.

6 If the conclusion of step 5 is not satisfying, then it might well be that the most important factor is not included in the set. So start again with a new set of factors.

Phase II: Elimination

7a Calculate the decision limits for the high/low level results (the medians of the high/low level results $\pm\overline{d}$ (compare step 3b from components search).

7b Run two experiments:

- □ The low level of factor A (A_L) and the high level of all remaining factors (R_H);

- The high level of factor A (A_H) and the low level of the rest (R_L).

8a No change (*i.e.*, $A_H R_L$ is within the decision limits of the low level results and $A_L R_H$ is within the limits of the high level results), indicates that A has no significant effect and therefore may be eliminated.

8b A partial change (*i.e.*, one result is beyond the decision limits) indicates that factor A is important, but only in combination with at least one other factor.

8c A complete reversal proves that A is the culprit factor, and that the remaining factors may be eliminated.

9 Repeat step 7b with the next important factor (B, C, ...) until the reversal is complete, or at least two important factors have been found.

Phase III: Confirmation

10 With two important factors (*e.g.*, A and B) run two new experiments, to check whether all remaining factors may be eliminated:

- The low level of the important factors and the high level of all remaining factors ($A_L B_L R_H$);
- Its counterpart ($A_H B_H R_L$).

Phase II has to be resumed with the next important factor when the reversal is not complete (step 7b); otherwise continue with the final phase (step 12).

11 If a third important factor has been found, then step 10 has to be repeated with three factors; this will not often be necessary, fortunately. Even rarer is the case that four factors or more are simultaneously important.

Phase IV: Analysis

12 Quantify the effects of the factors with a factorial analysis (see case illustration).

Case Illustration: Wire Sawing Silicon Wafers

In a diffusion centre silicon wafers are processed to give them specific semiconductor characteristics. Afterwards the wafers are separated into crystals, in this specific case by means of a sawing process. The yield, the number of usable crystals from a wafer, of this process varies enormously. An improvement team solved this problem with a variables search. They considered fifteen factors from four different categories:

- Sawing paste (four factors);
- Sawing wire (three factors);

- □ Machine parameters (six factors);
- □ Product (two factors).

All factors were ordered according to their perceived influence on the yield. The six most important factors are:

A Roughness of the sawing wire;

B Viscosity of the sawing paste;

C Carbon percentage of the paste;

D Wire tension of the machine;

E Age of the paste;

F Distribution of grain size of the paste.

The current setting of each factor was set as its low level; as its high level a perceived improved setting was chosen.

The first experiments had all factors on either their low level or their high level, in order to see whether the alternative was significantly better than the current situation. The yields were:

low levels:	85.7%, 87.1% and 91.4%	median = 87.1%	range = 5.7%
high levels:	96.7%, 98.8% and 98.1%	median = 98.1%	range = 2.1%

All high level yields are better than the low level yields, and the distance between the two medians ($D = 11.0\%$) is significantly larger than the average of both ranges ($\overline{d} = 3.9\%$): $D/\overline{d} = 2.82$ (> 1.25). The conditions of step 4 are therefore met. The decision limits are:

81.1% and 93.1% for the low level settings;

92.1% and 100% for the high level settings (larger than 100% is nonsense).

Then in the elimination and confirmation phases the following experiments were run (significant changes are printed in boldface):

$A_H R_L$: 87.4%	$A_L R_H$: 94.2%	A is not important
$B_H R_L$: **94.0%**	$B_L R_H$: 93.5%	B is important, but in combination
$C_H R_L$: 83.9%	$C_L R_H$: 97.9%	C is not important
$D_H R_L$: 86.6%	$D_L R_H$: 98.3%	D is not important
$E_H R_L$: 90.1%	$E_L R_H$: **88.8%**	E is important, but in combination
$B_H E_H R_L$: **96.4%**	$B_L E_L R_H$: **90.5%**	B and E are together decisive.

The factor F (and other factors) needed no investigation anymore, because the results were completely reversed. The variables search is summarized graphically in Figure 9.5.

The factorial analysis of the experiments is shown in Table 9.2.

The table shows that both viscosity and age of the sawing paste are important for the yield, both being about equally important. The combination of both factors (the interaction, see Figure 9.6) is more than the sum of the separate effects. The team decided to improve the viscosity of the paste, and to renew the paste more often in order to prevent it from becoming too old.

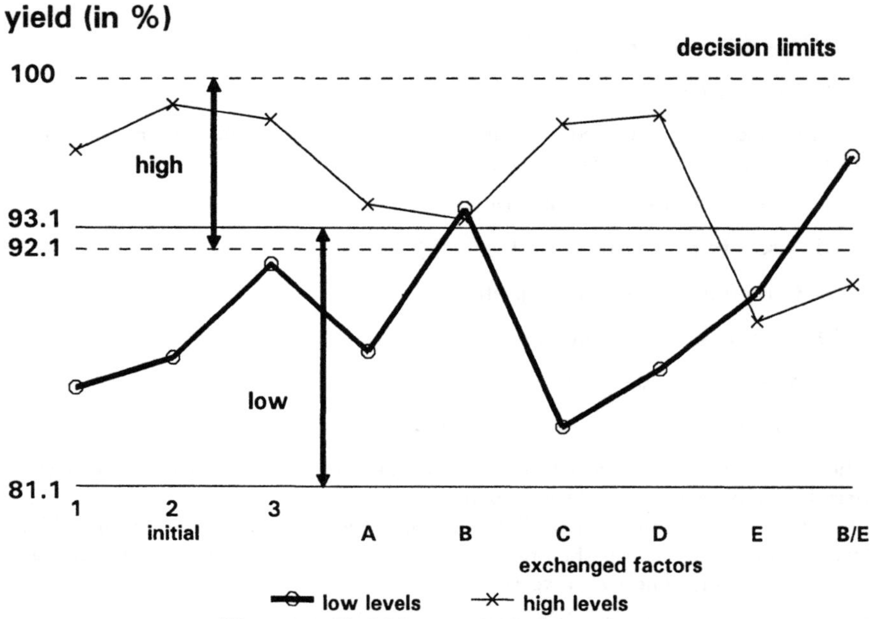

Figure 9.5: Variables search of wire sawing.

Table 9.2: Factorial analysis of wire sawing experiment

	B_L	B_H
E_L	85.7% 87.1% 91.4% 87.4% 83.9% 86.6% 90.5% Median = 87.1%	94.0% 88.8% Median = 91.4%
E_H	93.5% 90.1% Median = 91.8%	96.7% 98.8% 98.1% 94.2% 97.9% 98.3% 96.4% Median = 97.9%

9·6 Factorial Design

The preceding tools enabled us to limit the number of important factors to at most four. The factorial design allows us to pinpoint the culprit factor and possible accomplices, and to quantify their effects (interactions inclusive).

The power of a factorial design is that all possible combinations of low and high levels of the factors are tested. This allows us to distinguish main effects from

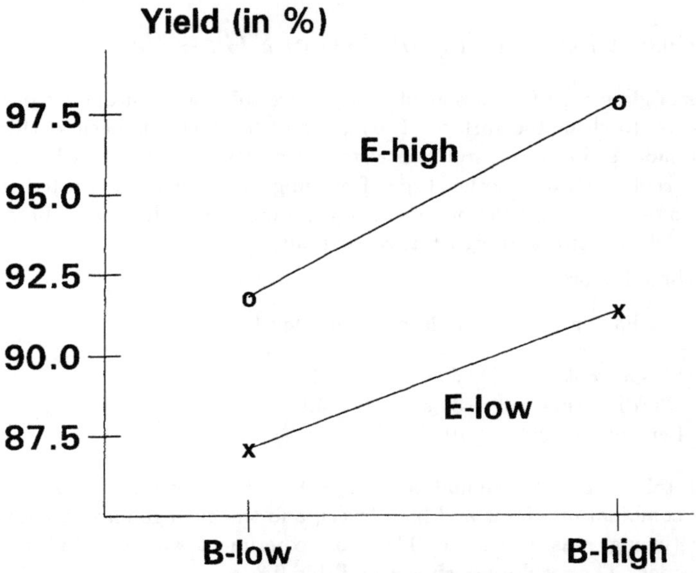

Figure 9.6: Interaction graph of factors B and E.

interaction effects (combined effects of two factors or more). For three important factors (*e.g.*, A, B and C) $2 \times 2 \times 2 = 2^3$ different combinations of levels are tested:

$$A_L B_L C_L, A_L B_L C_H, A_L B_H C_L, A_L B_H C_H, A_H B_L C_L, A_H B_L C_H, A_H B_H C_L, A_H B_H C_H.$$

For this reason this factorial design is also known as a 2^3 experiment. With an increasing number of factors the size of the factorial design increases exponentially. The number of factors is therefore limited to a maximum of four. A factorial design is an example of an experiment from a whole class of designs that has extensively been investigated (see *e.g.*, Box, Hunter and Hunter (1978), Duncan (1986), or Montgomery (1991)).

Factorial designs are also described in a number of steps.

1 Select at most four factors: A, B, C and D.

2 Select two levels for each factor, a low level (-1) for the worst results and a high level (+1) for the best results.

3 Write down all possible combinations of levels.

4 Arrange the combinations in a random order (use *e.g.*, a table with random numbers to designate the order).

5 Run a test for each combination.

6 Repeat steps 4 and 5 (the second order of tests is completely different from the first).

7 Compute the average of the two test results of each combination.

8 Compute the effects of factors and interactions (see case illustration).

Case Illustration: Pickling of Pens of a Glass Base

The pens of glass base (a component in a picture tube or monitor) are pickled for two reasons, to clean the surface of oxides, and to reach the desired thickness). Owing to more stringent environmental requirements the current pickling process had to be replaced with another type of pickling. The engineers started to determine the most important factors of the new process, and when only three critical ones were left a factorial design was carried out.

The three factors are:

Letter	Factor	(−1) level	(+1) level
A	Current	5 A	10 A
B	Pickling time	30 s	50 s
C	Percentage acid	10 %	30 %

The total number of combinations is $2^3 = 8$. A batch of glass bases was pickled for every combination of low and high levels, and the average mass decrease of 10 products (in mg) was measured. Then the experiment was repeated in another random order. The results are shown in Table 9.3.

Table 9.3: Measurements of the pickling experiment (mass decrease in mg).

	$A = -1$		$A = +1$	
	$B = -1$	$B = +1$	$B = -1$	$B = +1$
$C = -1$	10.7	18.6	22.6	31.9
	15.9	19.0	26.0	40.0
$C = +1$	17.9	33.1	43.7	59.5
	22.3	29.6	38.6	50.5

With the use of a calculation table (Table 9.4) the effects of the factors and interactions can be calculated. In this table all possible level combinations of the factors are written down, and the columns of interactions are added (simply by multiplying the columns of the corresponding factors). The last column contains the average of the two measurements of the cell. The last row contains the contributions of the factors and interactions: these are obtained simply by multiplication of the value in the column with the output, and then adding over the rows.

The contribution of factor A is computed as $-13.30 - 20.10 - 18.80 - 31.35 + 24.30 + 41.15 + 35.95 + 60.00 = 77.85$. The other contributions are computed similarly. The table shows that factor A (current) is the culprit factor, followed by factor C (percentage acid). Finally, factor B (pickling time) is an accomplice, and the interactions are not really important. All effects have a positive sign, meaning that the mass decrease is larger at the +1 level than at the −1 level.

Table 9.4: Calculation table for the pickling experiment.

A	*B*	*C*	*AB*	*AC*	*BC*	*ABC*	Output
−1	−1	−1	1	1	1	−1	13.30
−1	−1	1	1	−1	−1	1	20.10
−1	1	−1	−1	1	−1	1	18.80
−1	1	1	−1	−1	1	−1	31.35
1	−1	−1	−1	−1	1	1	24.30
1	−1	1	−1	1	−1	−1	41.15
1	1	−1	1	−1	−1	−1	35.95
1	1	1	1	1	1	1	60.00
77.85	47.25	60.25	13.75	21.55	12.95	1.45	

9·7 'B *vs* C'

This tool amounts to confirming our belief that a current process (or product, setting, ...) can be improved by a better process. In order to use this tool it is not necessary that the results can be measured, but ordering from good to bad should be possible.

B *vs* C is a tool which uses the fact of an alternative which really is better does score higher than the current situation. Consider the situation that two products from the 'better' process are compared with two products from the 'current' process. If the products (designated B and C resp.) are ranked in descending order, there are six possibilities:

1 C C B B

2 C B C B

3 C B B C

4 B C C B

5 B C B C

6 B B C C

The last possibility is the most likely result when the alternative is really better, whereas all possibilities are equally likely when there is no difference with the current situation. So, if in this experiment the result 'BBCC' is considered enough evidence that the alternative is really better, then elementary probability theory shows that the chances are only 1/6 that the alternative process is unjustly declared the better process. Smaller risk can be obtained when the sample size is larger (see Table 9.5).

B *vs* C can be described in five steps:

1 Select the maximal acceptable risk to declare the alternative process unjustly better than the current process. Choose from percentages 0.1 (super critical), 1, 5 and 10 (moderately critical). In industry 5% risk is common.

2 Select sample sizes according to Table 9.5. A common choice at 5% risk is three products from the 'better' process, and three products from the current process.

3 Sample the products in random order.

4 Arrange the products in order of decreasing quality.

5 If there is no overlap between the products (*i.e.*, all 'better' products are really better than all 'current' products) then we may conclude—with the selected risk—that the alternative process is better. Otherwise the two processes are not significantly different.

Table 9.5: Sample sizes for V *vs* C.

Risk	0.1%		1%		5%		10%	
	B	*C*	*B*	*C*	*B*	*C*	*B*	*C*
Sample sizes	2	43	3	13	1	19	1	9
	3	16	3	7	2	5	2	3
	4	10	4	5	3	3	3	2
	5	8						
	6	6						

B *vs* C is a powerful tool if the alternative is much better; small improvement will often not be discovered. Small improvement require much stronger statistical tools (such as Student's *t*-test) than B *vs* C, which is a non-parametric methods (a class of rather inefficient statistical tools).

Case Illustration: Contamination of Glass Tubes

Contaminated glass tubes may cause leakage currents in diodes, which, as a result, will be rejected. When the reject rate from contaminated glass was unacceptably high at a given moment, an improvement action together with the supplier of the glass tubes was started. The production of the glass tubes was modified, and to test whether the new process was really an improvement a B *vs* C experiment was agreed upon, and 1% risk was accepted.

During a period of a few weeks nine batches of glass tubes were tested, alternatively from the current process (four batches) and the modified process (five batches). The electrical yield of the corresponding batches of diodes are:

Current	87.2 %
Modified	95.1 %
Modified	93.2 %
Current	91.0 %

Modified	97.6 %
Modified	94.2 %
Current	78.3 %
Modified	92.8 %
Current	88.7 %

When the batches are ordered from high yield to low yield (BBBBBCCCC), we see that there is no overlap; all 'better' yields are higher than the 'current' yields. This proves with 99% certainty that the new process is indeed an improvement.

Chapter 10

From Control to Assurance

10·1 Introduction

For each process step the activity plan for SPC implementation ends with certification of the SPC point. The result of a PAT is evaluated in accordance with the assignment and with a particular standard (such as the checklist in Table 4.4). SPC implementation for this process step is then formally concluded. However, this does not mean that the personnel involved can sit back and relax. Without specific maintenance of the control charts and the out of control action plan, the results which have been achieved will ultimately be lost. To prevent this, in the last step of the activity plan agreement is reached concerning the way in which the SPC point is to be assured.

10·2 Assurance of an SPC Point by a Team

The good practice of getting a team to control a process step should be continued when assuring the SPC point. The PAT is followed up by a Process Improvement Team (PIT). Step 10 of the activity plan is therefore called 'from PAT to PIT'.

The participants in the PIT must have the same qualifications as those in the PAT. However, since it is advisable to involve as many people as possible in SPC, there are advantages in choosing different operators for the PIT. Participation of the supporting disciplines is also less self-evident. For each PIT it will be necessary to see which support should form a permanent part of the team. For example, if the process performance is high enough, input from anT:R&D specialist will not be of any additional value. And if the problem solving skills of the team members are sufficiently developed, the role of the SPC expert will be limited. Participation of the process engineer remains necessary (as person responsible for the technical functioning of the process), but in many organizations his role can be smaller than in the PAT. To emphasize the role of direct production the supervisor, or perhaps even an operator, should be chairman of the PIT. It is important that the PIT is seen as a permanent part of the production organization which, however, leaves the formal structures intact.

10·3 Assignments for a Team

In the first place the Process Improvement Team is responsible for maintenance of the control charts and the out of control action plan of the SPC point. Seeing that all operators contribute collectively to controlling the SPC point, the team cannot be given sole responsibility for this. However, it does have the task of charting the uncontrolled elements, signalling trends in uncontrollability and suggesting possible remedies. With this, the PIT begins on its second task: improving the process step where, in the first place, the out of control action plan is the most appropriate starting point. Sandorf and Bassett (1993) describe their practical experience with dynamic out of control action plans.

Not only the SPC point is eligible for improvement. Various other aspects of the process step can also be improved. For example:

- safety and tidiness in the work place;
- ergonomic improvements;
- reduction of waste (waste, refuse, waiting periods,...);
- optimization of the logistics (throughput times, intermediary stock, ...).

Incidentally, extension of the assignment to this type of subject is not at all revolutionary, since the relationship with control is obvious. However, for managers who associate SPC with control charts this step may be one too far!

There is, however, another reason why extension of the assignment may be desirable. Many PITs will find the assignment to assure and improve the SPC point an easy one. If the process is properly controlled, meets the demands, and contains no surprises, the challenge for the team will be limited. The chances are that in time attention will wane and that therefore control will also diminish, besides the fact that stagnation also means a step backwards. Furthermore, in this situation the potential of the operators to initiate improvements is sadly unused. Later in this chapter we will discuss the organizational form which best suits improvement teams, and their management (shop floor management).

10·4 Modes of Operation of a Team

By examining the results of the out of control plan in a Pareto analysis, the most important causes of uncontrolled situations will immediately become evident. If the team can come up with solutions for this, the process variation diminishes and process performance increases. For example, wear of tools turns out to be the most important cause of warning signals in the control chart, despite that they should be replaced regularly as a preventative measure. The team can then examine various aspects:

- ☐ Are the tools replaced regularly as a preventative measure?
- ☐ Is there much variation in the quality of the replaced tools?
- ☐ How long does it take for tools to become worn?
- ☐ Is there variation in rate at which the different tools become worn?
- ☐ How large is the variation between the different new tools?
- ☐ Is it possible to measure the wear? If so, how great is this?
- ☐ Does it help if the tools are cleaned (more thoroughly)?
- ☐ Can the tools be made of better wear resistant material?
- ☐ What is the cause of wear, and can this cause perhaps be reduced or removed?
- ☐ Is it possible to replace worn tools more easily?

To investigate some of these questions a measuring point will have to be set up, others require an experiment, others still are of a more theoretical nature (*e.g.*, material science) etcetera.

The team can use the first five steps of the activity plan in Chapter 5 for assignments which have no direct connection with the SPC point. Suppose that the team is given the assignment to reduce the amount of intermediary stock after a process step. They could tackle this problem as follows:

- ☐ Make a detailed description of the logistic process (step 1);
- ☐ Consider what the most important causes of stock-forming are (step 2);
- ☐ Classify the causes in order of importance (step 3);
- ☐ Take action for improvement (step 4);
- ☐ Set up one or more measurement points (step 5).

This model can be applied to many problems. With it, however, the (possible) improvements are not yet fact. According to the PDCA cycle of continuous improvement, the effectiveness of the improvements must first be tested before their possible implementation as a standard.

The team will find it easier to think of improvements if the participants are aware of the problem solving tools in the previous chapters.

10·5 Progress Control of the Team

Specific assignments to the team are naturally monitored by the person who has commissioned them. But the basic assignment (assurance of the SPC point) must also be monitored periodically. The certificate in step 10 has therefore only a limited validity, *e.g.*, one year. The process step must then be certified once more. In this way the Steering Committee makes a formal check once a year whether or not the process step is still controlled and in accordance with the requirements.

The annual re-certification provides the Steering Committee with an important means to start the cycle of continuous improvement. The team can be left free to choose the subjects on which it is going to work. One or more target figures can be indicated for the coming year (*e.g.*, , increase C_{pk} to 1.5, halve the number of uncontrolled situations, reduce the number of accidents to 0). The assignments must, of course, be embedded in the department policy.

10·6 An Organization for Improvement

In this section an organizational form in which process improvement teams can work well is described. The picture given is more or less the ideal situation, a stage not (yet) reached by many companies. Furthermore, in some organizations the culture is such that the picture sketched will certainly not be sought after. ASM Lithography is a good example. This company is so strongly oriented to development that production will never be able to play the role we sketch here.

So far we have written little about the organizational form of the company or institution where SPC is being implemented. In Chapter 2 we indicated a number of conditions which an organization must meet in order to be able to begin implementing SPC with any chance of success. And in Chapter 3 we described the role of the management and the project-based character of SPC implementation (with a steering committee and Process Action Teams, PATs). The idea behind this is that, in principle, any organization, industrial or otherwise, can successfully make SPC an essential part of operations management. However, we must acknowledge that an organization which has concluded an SPC implementation project successfully, has probably changed considerably during the project. Some achievements are that personnel have:

- Had tasks, responsibilities and qualifications to improve their own place of work;
- Learned to defend measures and actions against criticism by their colleagues;
- Learned to work in multi-disciplinary teams;
- Presented their achievements in public.

These achievements can perhaps best be summarized by stating that a large number of people within the organization has become more independent. Nowadays this is also termed 'empowerment'.

The chances are that the distribution of tasks, responsibilities and qualifications, the supporting systems and the management of the organization no longer fit together after SPC implementation. The organization may have become a straitjacket, impeding continuous improvement. Adjustment to the new situation is necessary. A possible organizational form which suits the new situation is one based on socio-technical principles, with self-guiding teams (also called production cells, clusters, autonomous groups, whole task groups, *etc.*). In Frame 10.1, the characteristics of such an organization and the draft rules which must lead to it are given (cf. Haak (1994)).

Every organization will be able to split up the process stream into a limited number of elementary steps (which form the basis) and a large number of supplementary steps and auxiliary treatments. The elementary steps can be characterized by the fact that their added value is associated with the product, that they are irreversible and that they have a high risk of failure. The socio-technical organization is built up around these elementary steps. With SPC, on the other hand, a form of control has been installed around each elementary step. Each autonomous group is responsible for at least one elementary process step and accommodates at least one Process Improvement Team. To ensure no step is excluded it is advisable to bring every process step under the care of a PIT.

An autonomous group which is not too large (a maximum of about 12 persons) can coincide with a PIT. In larger autonomous groups, the PIT is a representative team with a special improvement task. It is possible to choose for regular changes in the composition of the PIT, so that everyone in the autonomous team participates now and then. Or various PITs can be set up, so that everyone in the group participates in a team, and each PIT has a separate improvement task. In this last model it is necessary that (a few of) the operators are competent in improvement while working in teams. Training will be necessary here. Of course, people differ, and not everyone is necessarily suited or motivated enough to carry out an improvement task.

For the management of such an organization, the principles of shop floor management can be applied successfully.

10·7 Shop Floor Management

Suzaki's 'shop floor management' philosophy (1987 and 1993) places the people in the organization in a central place. The management style entails that all employees must be convinced that their own individual development also means development of the organization. In order to develop, the personnel must be given self-regulating (or autonomous) opportunities which are related to the objectives of the whole organization. The theme of shop floor management is that waste, in

Frame 10.1 An Organisation with Autonomous Groups

Nine principles for an organization with autonomous groups are:

1 The task of the group must be complete, must have clear boundaries and must be coupled to a quantifiable result.

2 The group must have sufficient opportunities for adjustment so that it can carry out the task as independently as possible.

3 The tasks of the members of the group must show interdependency, as their activities complement each other.

4 The size of the group must be such that it can make an identifiable contribution to the organization, that it can make quick decisions and that it is not too vulnerable.

5 The members of the group can be set in for various tasks within the group. Internal differences in status should hamper neither a flexible distribution of work nor internal mobility within the group.

6 There must be a contact person within the group both for the external world and for the group members themselves.

7 The group must have its own room(s), have access to its own production resources and to information.

8 The administration and management systems must be in keeping with the independence and responsibility of the group.

9 The reward system must be in keeping with teamwork.

For designing such an organization one needs:

1 The layout of the organization is a derivative of developments and requirements from the external world.

2 The organization is designed in the following order: first the primary process, then design of the process management and finally design of the systems (budgeting, planning, procedures, *etc.*).

3 The production structure is drawn up proceeding from the whole to the separate parts.

4 The management structure is drawn up proceeding from the separate parts to the whole.

all its forms, must be fought against. The creativity of personnel is used in order to locate waste and eliminate it, starting from the reality of:

- □ 'Genba'—the actual place of work (where it all happens),
- □ 'Genbutsu'—actual things (products, materials, machines),
- □ 'Genjitsu'—actual facts (correct data, causes).

In addition, the organization must be customer-oriented. Those organizations which best fulfil the customer's wishes will win from the competition. It is, therefore, essential that everyone within the organization is aware of his customer–supplier relations (both inside and outside the organization). This goes for all aspects of the relationship: materials, information and services. Each employee must constantly consider the problem of how to provide his customers with better service, thus continuously improving himself and the organization. By means of communication, coordination and cooperation, care must be taken that personal interests do not prevail above the objectives of the total organization. 'Kanban' (Suzaki, 1987) can be a useful tool here.

Entrepreneurship and customer orientation are given shape in the form of a mini company, a company within a company (see Figure 10.1).

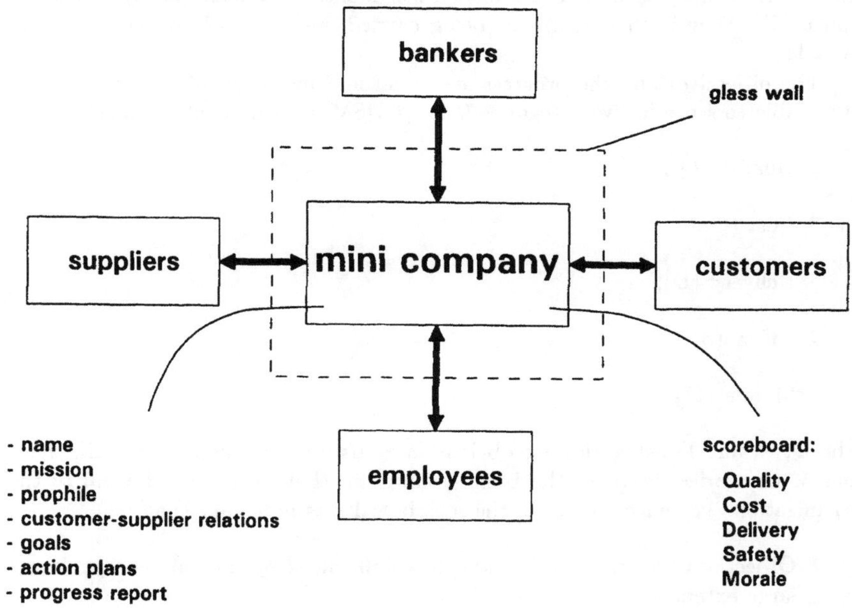

Figure 10.1: The Mini Company.

Under the leadership of a director the mini company:

- □ Develops a name and a mission;
- □ Draws up a profile and takes stock of customer–supplier relationships;
- □ Formulates goals and draws up an activity plan to achieve these goals;

- Carries out the plan;
- Makes progress reports.

This cycle is repeated regularly.

The mini company has relations with four parties:

- With the bank (which makes means available);
- With customers (who purchase products, services or materials);
- With suppliers (who offer products, services or materials);
- With personnel (employees or machines, who carry out the tasks).

A mini company is characterized by open communication with all parties. In shop floor management this is called glass wall management, as opposed to stone wall management. Its distinguishing characteristic is that the work place is organized in such a way that the answer to any normal question ('Where is that product going to?', 'Are these faulty products?', 'Are you ahead of planning?', 'How many ideas were submitted last month?', 'Which quality control do you use?', 'Why is this treatment being carried out?') is self-evident, even for outsiders.

The objectives and the progress are monitored by means of a scoreboard. It is possible to score in five categories (the QCDSM measurements), namely:

- Quality (Q);
- Cost (C);
- Delivery (D);
- Safety (S);
- Morale (M).

The Q, C, and D categories are obvious in a customer oriented approach, the S and M categories interpret the basic assumption that people are central to the organization. For effective use of the scoreboard it is necessary that

1 Objectives are measurable and can be influenced by the mini company to a large extent;

2 The number of main objectives is manageable;

3 The specific objectives are coupled to the objectives of the whole organization;

4 People have access to resources and support enabling them to work on the objectives.

With the above the most important characteristics have been described. For further interpretation and the means used for it, we refer to Suzaki (1987 and 1993).

In a socio-technical organization, it is to be expected that autonomous groups will function according to the mini company concept. Within the mini companies, PITs could be given special improvement assignments for one or more of the QCDSM objectives. Other objectives of the mini company could be carried out by other PITs. The PAT notice board (see Frame 4.3) is a partial interpretation of the scoreboard.

Chapter 11

Software for SPC

11·1 Introduction

Managers who wish to implement SPC are often under the impression that computerization is necessary for this purpose. There is thus a lot of SPC software on the market. Renowned statistics software packages offer extra modules for control charts, process performance analysis, Ishikawa diagrams, and flow charts. The risk here, however, is the assumption that by implementing such software SPC has also been implemented. This is certainly not the case.

For most of the activities in the approach presented in this book it is possible to require no more than pen and paper, a good word processor, a spreadsheet programme with statistical features, and software for making flow diagrams. The data necessary for setting up a control loop are usually limited, and the analysis is simple. At the outset of implementation it is advisable to start in this manner, for two reasons:

1 The focus is on becoming familiar with the operating procedure and the methods, and is not diverted by the complex problems which ensue from the implementation of software;

2 By gaining experience with the methods by means of pen and paper, it will become evident exactly which computerization needs exist within the organization.

On the basis of experience thus gained, software for SPC can be purposefully and efficiently implemented where necessary, and integrated with existing computer systems. In this chapter we present a framework for the application of software for SPC. This framework can be used as an aid for choosing software, deciding where this will be implemented and where not, and possible integration with other systems. Given the speed of development in this area, specific software will not be presented and evaluated. The guidelines in this chapter will provide sufficient information to evaluate and make a choice oneself.

11·2 A Framework for SPC and Software

When SPC has been implemented on the shop floor there are various levels at which software can support the activities. The following functions can be recognised:

1 Performing a control loop on a process step by operators on the shop floor.

2 Evaluation of the performance of the control loops by supervisors or production managers (see the checklist for maintenance of an SPC point in Table 4.4). This goes both for the separate SPC points as well as for the system of control loops for the whole process.

3 Evaluation of the results of a process step (process performance, out of control plan signals, logbook data, *etc.*) as a basis for improvement.

4 Evaluation of product–process relationships to achieve improvement in the whole process.

This is shown schematically in Figure 11.1.

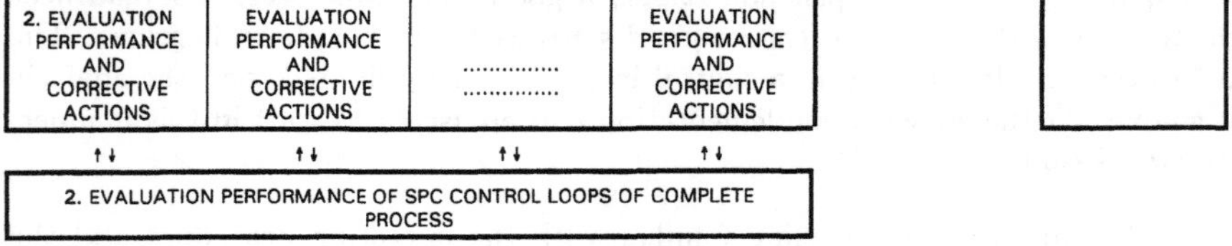

Figure 11.1: Outline of functions in SPC points.

Each level has specific computerization requirements. If SPC has been properly embedded in the organization, software will be fully integrated and coupled to the logistic management. In most cases, however, it will be possible for quite a while to work well with partial solutions in each area. In this case it is important to save the data in such a manner that they are identical for all applications (no double storage), and that product characteristics can be traced back to the state of the various process steps when the product was produced (for evaluation of the relationship product process). Traceability of products is essential here (and also a requirement for ISO 9000).

In the outline in Figure 11.1 the following considerations apply:

1 SPC point: Process step with control loop.
An SPC point on the shop floor is characterized by frequent transactions: input of random check results, checking control chart, signalling, taking a look at out of control plan, intervention and registration of the transaction. Computerization of these transactions (with the exception of intervention, of course) has the following advantages:

- Data is registered immediately;
- Software can ensure that intervention is always registered;
- More complex control charts can be used;
- Information about the functioning of the control loop and the results of this in the process are easily accessible.

An extremely fast response is required of software for this part of SPC, since the transactions are dispatched on-line. In principle this must be possible, in view of the simplicity of the transactions. Often, however, it proves to be a problem because of integration with logistic operations software. A response time which is too long makes the system impractical.

2 Performance of the control loops.
Important information for evaluating the performance of the control loops will contain the following elements (if the control chart and decision rules are given by the software):

- correct input of data;
- correct application of the out of control plan;
- correct following up of work instructions.

These aspects cannot be completely monitored by means of software registration. A good method is to make a review of the last period on the basis of the registered information. The supervisor can discuss this review with the operators concerned to see if the realization is taking place according to plan. In the same manner all SPC points within a process can be periodically checked. Structural problems in the handling of control charts and out of control plans can be discussed on the basis of the results. These actions are typically off-line, and a fast response is less important. Well structured data and simple software to generate the overviews/reviews are, however, essential.

3 Evaluation of the result and process improvement.
Information necessary for the evaluation of the result of the process will contain the following elements:

- the number of times the process was 'out of control';
- which solutions given in the out of control plan were used;
- which actions were taken;
- maintenance reports;
- inherent process variation during the last period.

On the basis of this information, the PIT can periodically adjust the control limits of the out of control plan. Also, small improvements can be made on the basis of this information. This is, again, an off-line transaction, so that as far as software support is concerned the same requirements as above are valid: a good, accessible data structure and user friendly software for making reviews and analyses. Statistical software can be of help here, especially in view of this last point—the variation inherent to the process—and the investigation of possible improvements.

4 Process–product relationships.
Statistical software is indispensable for the investigation of process–product relationships. More advanced statistical methods are also necessary for more fundamental improvements to a process step to reduce the inherent process variation. Crucial requirements are well structured data, traceability of products throughout the whole process, and user friendly software.

11·3 SPC Software in Practice

There is a gigantic amount of SPC software. The possibilities of the packages varies greatly: from small PC-packages for control charts and C_{pk}'s to large packages for mainframes and work stations with modules for application in all areas of statistics. This last category usually also offers complete menu systems to serve various areas of SPC. The most sensible strategy for implementing SPC software depends on a number of limiting conditions within the organization:

- □ What kind of software support is needed?
- □ What kind of infrastructure is already present (computers, networks, data storage, traceability)?
- □ How 'computer minded' is the organization?

To discover the critical points in these conditions, and so make a purposeful choice for software, a number of phases are passed through. These phases are parallel to those passed through in SPC implementation (we follow the phasing in Chapter 3).

Phase 1. Awareness

SPC software plays no role in this phase, with awareness days for all concerned.

Phase 2. Pilot Projects

Introduction of control charts and out of control plans on the shop floor with pen and paper or a simple spreadsheet application. The necessary data for this are also collected by hand. Analysis takes place by means of a spreadsheet, or a statistical package—which is easy to use on a PC, in general. The statistical package is usually not specifically SPC-oriented, since use is made of other statistical methods in the analysis of data. The requirements for data and data processing will become clear on the basis of information acquired in this phase.

Phase 3. Integral implementation

In this phase it is very important to establish a good data structure and (if still necessary) to ensure traceability of products. This can be software especially developed for this purpose. Given the necessary response performance and the relative simplicity of the transactions, a statistical package is not necessary, and even sometimes not advisable. The solution will depend strongly on the software already in use in the organization. The off-line SPC activities will use the same data, with the help of a spreadsheet and a user friendly statistical package.

Phase 4. Total quality by means of SPC

In the meantime the requirements for the off-line activities will have become clear. Problems in the data structure and the infrastructure have been recognized and are being solved. The 'computer mindedness' of the organization has been more than tested. This is an appropriate moment. to implement integral solutions if these are necessary, given the size of the company. At this point the requirements to which this software must comply can be formulated more clearly. An important point to keep in mind is that the cost of the software itself will probably be the smallest cost. Implementation of this in the hardware environment and training of personnel is more expensive, not to mention the cost entailed in making the wrong choice!

11·4 Does Software Make SPC More Attractive?

From what has just been argued above it will be apparent that one should not think lightly about implementing various software tools for SPC. The most important reasons for this are:

1 Attention must be fixed on becoming familiar with the operating procedures and methods;

2 Using pen and paper, experience with the methods is obtained. In this way, the existing computerization needs will be discovered.

Further, the visibility of the on-line SPC activities on the shop floor have an important positive influence on motivation. SPC is often less visible when fully computerized.

Nevertheless, our conclusion is: software can make SPC a lot more attractive. It lightens work, jazzes up presentation and opens up (off-line) possibilities for analyses which can lead to improvement. Experimentation with data becomes easier and resistance to keeping various lists up to date clearly diminishes. So, as long as it is clear where exactly it should be set in, use of software can provide a positive contribution to the implementation of SPC.

Chapter 12

SPC Competition and Self-Evaluation

12·1 Competition for the SPC Team of the Year

In section 3.3 we stated that it was one of the tasks of the SPC Steering Committee to develop initiatives for promotion. One of the most effective forms of promotion is an annual competition of the best SPC team. Before such a competition can be held, however, it will be necessary for the company to have already made some headway with the implementation of SPC. A suitable time to organize a competition would be about one year after the start of SPC implementation. The competition offers PATs and PITs the opportunity to present themselves and their results to the management and the other personnel. The reason for a competition is that the accompanying suspense heightens the commitment of the participants, other employees, and the management. The competition also meets the needs of the team members for recognition and appreciation. Publicity centred around the competition must be well taken care of, as it is one of the moments that SPC is put in the spotlight within the company. The existing means of communication in the company (placards, billboards, department paper, company paper, SPC bulletin or suchlike) can be used to announce the competitions, and to report on them.

The SPC Steering Committee is responsible for the organization of the competition. Depending on the number of teams who wish to participate, it may be necessary to have preliminary rounds for the competition. These preliminary rounds are identical to the final round, and do not need to be described separately. Amongst other things, the Steering Committee must:

- □ appoint a jury;

- □ draw up the assessment criteria;
- □ organize registration of the teams which have enlisted for the competition;
- □ provide a suitable hall with all the necessary facilities for ensuring good presentation;
- □ coordinate the publicity around the competition;
- □ organize prizes (which are in accordance with the importance given to SPC by the management).

Most of these points are self-evident. We would like to elucidate two of the points: the jury and the assessment criteria. The chairman of the jury is a member of the management team (General Manager or Manager Operations) and the members are *e.g.*, the Quality Manager (because of his specific knowledge in the areas of quality control and quality improvement), the HRM Manager and a representative of the employees.

Our assessment list is based on the 'Quality Improvement Competition' (QIC) of Philips' Semiconductors Product Division. In 1993 this product division started a competition for the improvement team of the year. The QIC is a multistage competition: local competitions are held first, the winners of which are delegated to the regional competitions. Finally, the world final is held, which ultimately yields the improvement team of the year.

The assessment criteria are related to the following subjects: mutual cooperation, manner of operation, analysis of the problems, solutions, results, impact on the organization and the presentation. Besides evaluating the financial results, the reason various perspectives are considered is, *e.g.*, to ascertain the capacity of the team to solve problems. The assessment criteria (see also Table 12.1) are divided into seven categories which can be awarded a maximum of 10 or 20 points, depending on the importance attached to them. Each category contains a number of questions to examine to which a score should be given. The jury's task is to assess the presentations of the various teams by awarding points with the help of the score list. Thirty minutes is available for each presentation, after which the jury may ask questions if points need clarification. The team with the most points wins (note that the jury has a thankless task: in the eyes of the losing teams they have done a bad job!)

Table 12.1 Assessment Criteria for the SPC Competition

The complete list of assessment criteria and the maximum amount of points which can be achieved:

1 Teamwork (maximum of 10 points)

- 1.1 Were meetings held regularly?
- 1.2 Was the attendance at these meeting by team members sufficient?
- 1.3 Were all team members involved in decisions?
- 1.4 Were the individual skills utilized?

1.5 Are there indications of a good team spirit?

2 Problem definition (maximum of 10 points)

2.1 Have the internal and external customers been taken into account?

2.2 Has the process been described well?

2.3 Are the goals for improvement clearly defined?

2.4 Are the goals for improvement quantified?

2.5 Are the goals for improvement realistic?

3 Process analysis (maximum of 20 points)

3.1 Has the PDCA cycle been applied?

3.2 Have priorities been set?

3.3 Has the effect on other processes been taken into account?

3.4 Have the following problem solving methods been used: histogram, Pareto analysis, cause and effect diagram, scatter diagram or other quantitative methods?

3.5 Has the way in which a solution has been reached been clearly documented?

4 Solutions (maximum of 20 points)

4.1 Are the solutions related to the goals for improvement?

4.2 Are the solutions efficient?

4.3 Has an implementation plan been used?

4.4 Has there been sufficient communication about the changes?

4.5 Have the changes been implemented in a sufficiently creative (innovative) way?

5 Results (maximum of 20 points)

5.1 Are the improvements correlated to the goals for improvement?

5.2 Have the improvements been quantified and compared with the starting situation?

5.3 Are the improvements also noticeable?

5.4 Which goals for improvement have not yet been realized, and why?

6 Impact and assurance (maximum of 10 points)

6.1 Are the improvements permanent?

6.2 Can the results be used by others within the organization?

6.3 Will suppliers profit from the new approach?

6.4 Has the team created a climate for continuous improvement?

6.5 Have the results been properly assured?

7 Presentation (maximum of 10 points)

7.1 Has the whole team contributed to the presentation?

7.2 Do the tables and figures give enough information?

7.3 Does the presentation demonstrate that an improvement procedure has been followed?

12·2 SPC Self-Evaluation

As a final subject in this book we would like to hold a mirror up to companies which have been seriously involved in SPC implementation, to show to what extent this implementation has been successful. In this section we present a list for SPC self-evaluation for determining how far the business has come.

In large American multi-nationals such as Ford, General Motors, Chrysler, Lockheed Martin, Boeing, Motorola and IBM, it is general practice to include suppliers in their own SPC programme. One of the requirements for becoming (and remaining) a supplier is that SPC has been (or is being) implemented. The American companies are very helpful with the implementation. Complete handbooks for simplifying the procedure are often put at the disposal of suppliers. At the companies where we have assisted in implementing SPC we have obtained the following experience:

- At Philips Semiconductors, Stadskanaal, we had to deal with the 'Total Quality Excellence' (TQE) programme and 'Quality System 9000' (QS-9000) of the American automotive industry.

- At Hollandse Signaalapparaten we had to take the requirements of the aircraft builder Lockheed Martin into account.

- At ASM Lithography we were confronted with the fact that American customers in the semiconductor industry insist that SPC has been implemented.

Large American companies also have the habit of indicating exactly those points by which a supplier will be assessed. A checklist is made available at the outset with which the company concerned can evaluate how far it has been successful in implementing SPC in the manner required. This is, of course, only reasonable. If SPC implementation is one of the requirements of becoming a supplier, it is reasonable to indicate the manner in which the customer will grade its satisfaction with the implementation. This self-evaluation and the criteria used consequently lead to clarity for both parties.

We have assembled a self-evaluation checklist from the lists of Lockheed Martin, Ford and the Institute for Interconnecting and Packaging Electronic Circuits. This list can be used not only for measuring progress, but also for determining the extent to which the company's suppliers have been successful in SPC implementation. It is possible to certify suppliers on the basis of this checklist. The advantage of this is that the company acknowledges complete trust in the supplier, so that, *e.g.*, inspection of goods from this supplier can be omitted in the future. It is a step on the way to a sound relationship or partnership.

Before presenting the SPC self-evaluation checklist, it is important to qualify it somewhat. Firstly, we do not wish to suggest that the checklist is exhaustive. Additions or deletions can be made, depending on the type of organization. Secondly, the content of this book has been used as the line of approach. The terminology is geared to it, and we also concentrate on SPC implementation in a production environment.

The internal or external auditor who uses this checklist in his assessment of the extent to which a company has been successful in SPC implementation should, of course, check how far the answers given agree with reality.

SPC Self-Evaluation Checklist

0 **General Information:**

Comment: When evaluating the extent to which a company has succeeded in implementing SPC, demands are obviously made of the quality system. In this self-evaluation checklist we have restricted ourselves. In principle we assume that the quality system is in agreement with one of the ISO norms. For these companies we assume that, e.g., *procedures, working methods, and handling of complaints are properly documented. resp. organized. This does not necessarily mean that the company must be certified. The quality system must be set up in such a manner that the ISO norms are met. The book by the Bureau of Business Practice (1992) is warmly recommended to those companies where the quality system has not yet been fully developed.*

Date:

Name of company:

Department:

Address:

ISO certification:

If so, which certificate:

Since when:

If not, give a short description of the present quality system, referring to the various sections of the relevant ISO norm:

Person ultimately responsible for SPC:

Name:

Position:

Contact person:

Name:

Position:

Number of employees:

Annual turnover:

Type of business:

Description of the product portfolio:

1 **Management Involvement:**
Comment: These questions are designed to determine the degree of involvement and participation in SPC of the various management levels in the company.

1.1 Name the management of your company:
Executive Director:
Director:
General Manager:
Manager:

1.2 Has your company a written SPC strategy which has been communicated to all employees?
□ yes □ no
If yes, how? If not, why not?

1.3 On average, how much time does the management of your company spend on visiting and showing interest in the teams which are implementing SPC?

□ 0 hours

□ Less than 3 hours per month

□ 3 to 6 hours per month

□ More than 6 hours per month

1.4 During the past two years how much time has the management of your company, on average, spent on participating in training in SPC?

- ☐ 0 hours
- ☐ 1 to 8 hours
- ☐ 9–24 hours
- ☐ 25–40 hours
- ☐ More than 40 hours

1.5 How often does the management of your company require feedback on the progress of the various SPC activities?
Give a short clarification!

- ☐ Daily
- ☐ Weekly
- ☐ Monthly
- ☐ Quarterly
- ☐ Yearly
- ☐ Never

1.6 Does the executive management stimulate SPC enough to ensure that the middle management strongly supports SPC?
☐ yes ☐ no
If yes, how? If no, why not?

1.7 Has the executive management made all necessary resources and facilities available so that the implementation of SPC in the company is possible?
☐ yes ☐ no
If yes, how? If not, why not?

2 **SPC Coordinator and SPC Steering Committee:**
Comment: An SPC Steering Committee is needed to mark out the implementation course, to guide, and, where necessary, make corrections (see section 3.3 for a complete list of tasks). The SPC coordinator is the person within the company with the greatest knowledge of SPC. He lends the SPC Steering Committee a helping hand, functions as a walking encyclopedia within the company in the field of SPC, and generally supports the various SPC activities.

2.1 How many persons directly support the SPC implementation?

- ☐ 1
- ☐ 2–3
- ☐ 4–5
- ☐ 6–7
- ☐ 8–9
- ☐ 10 or more

2.2 Which positions and groups are present in your company?

- ☐ SPC Steering Committee (who participates in this group?)
- ☐ SPC coordinator (who?)
- ☐ SPC teams (which?)

2.3 At what level is the training for the supporting positions?
Tick the most relevant possibility/possibilities and give a short clarification!

- ☐ Self tuition
- ☐ On the job training
- ☐ 4–8 hour internal course
- ☐ 2–5 day internal course
- ☐ 2–3 day external course

2.4 In which manner is the progress of SPC charted within the various departments?
Check all relevant answers! Give examples!

- ☐ Reports on the SPC projects to the management
- ☐ Keeping track of the financial profits of the improvements
- ☐ Checking the extent to which employees are involved in implementation
- ☐ Keeping track of the number of SPC applications
- ☐ Other

2.5 How much time does the SPC coordinator spend on SPC per week?

- ☐ Less than one day
- ☐ 1–3 days
- ☐ 4–5 days

3 **SPC Procedures:**
Comment: It is important that the work methodology and the methods used are clear, consistent, and uniform. This prevents problems and discussion. This also should be aimed at for the use of statistical methods and techniques.

3.1 Are the various tasks, qualifications and responsibilities recorded?
☐ yes ☐ no
If yes, how? If not, why not?

3.2 Do operational procedures exist for the following points?
Check each point which has been regulated and give a short explanation!

- ☐ Carrying out a Gage R&R analysis of the measurements

- □ Setting up control charts and calculating the control limits
- □ Reacting adequately to signals in the control charts
- □ Setting up an out of control action plan (OCAP)
- □ Calculating C_p and C_{pk}
- □ Carrying out Process Capability Studies (PCS)

3.3 Have all employees who are involved in SPC implementation been given an explanation of the procedures and instructions?
□ yes □ no
If yes, how? If not, why not?

3.4 Are the most important customers aware of the latest version of the SPC procedures?
□ yes no
If yes, how? If not, why not?

4 **SPC Training:**
Comment: The importance of training has already been strongly emphasized by Deming in points 6 and 13 of his management theory. SPC will not take off without thorough instruction.

4.1 Does your SPC education and training programme contain the following ingredients:

Check all relevant points and give a short explanation!

- □ Compulsory training of all employees involved in SPC
- □ List of attendance at the training
- □ Examination of what has been learned
- □ Certification of the participants
- □ Re-certification of the participants
- □ Retraining programme

4.2 What percentage of the employees in the following categories has received SPC training?

	none	1–25%	26–50%	51–75%	87–100%
Executive management					
Management					
Technical staff					
Operators					

5 **Process and Product Characteristics:**
Comment: It is of great importance to keep track of the proper quality characteristics. It must be ensured that a structured method is used for carrying out this task.

5.1 Have all procedures of all process steps been described and filed away?
□ yes □ no
If yes, how? If not, why not?

5.2 How are quality characteristics selected?
Check the relevant answers!

- ☐ Customer wishes
- ☐ On the advice of the suppliers
- ☐ Waste registration
- ☐ FMEA
- ☐ Other

5.3 Who is responsible for selecting quality characteristics?
Check the relevant answers and indicate what the arrangements are!

- ☐ Quality Department
- ☐ Research & Development
- ☐ Production
- ☐ SPC teams
- ☐ Other

5.4 Where are the quality characteristics documented?
Give a short explanation!

- ☐ Nowhere
- ☐ With the work regulations
- ☐ In the Planning department
- ☐ In diagrams
- ☐ In the control plan (see section 3.3)
- ☐ Other

6 **Status of the Processes:**
Comment: The aim of SPC is to create a climate of continuous improvement. This can be achieved by reducing variation. Accommodating customer wishes should be taken into consideration when doing so.

6.1 What percentage of the process steps has been evaluated by means of a PCS?

- ☐ 0%
- ☐ 1–25%
- ☐ 26–60%
- ☐ 51–75%
- ☐ 76–100%

N.B.: A minimum percentage of 51% must be attained in order to be able to state that a company has made reasonable progress with SPC. There should be a blueprint as to how 100% can ultimately be realized.

6.2 For the processes for which a PCS has been carried out: what percentage has either a C_{pk} of at least 1.33 or a yield of at least 99.9%?

- ☐ 0%
- ☐ 1–25%
- ☐ 26–50%
- ☐ 51–75%
- ☐ 76–100%

6.3 Does an improvement plan exist for attaining higher C_{pk}'s or yields?

- ☐ yes
- ☐ no
- ☐ not applicable

If yes, how? If not, why not? If not applicable, explain.

6.4 How regularly are PCSs carried out?

- ☐ No rules
- ☐ Quarterly (every three months)
- ☐ Twice a year (every six months)
- ☐ Annually
- ☐ Other: explain!

6.5 Which part of the company uses the results of the PCS?
Check the relevant answers and give a short explanation!

- ☐ Production, to improve the processes
- ☐ R&D, to improve the design
- ☐ R&D, to revise the specifications

6.6 If only attributive data is available, is there an improvement plan for realizing a yield of 99.9%.

- ☐ yes
- ☐ no
- ☐ not applicable

If yes, how? If not, why not? If not applicable, explain.

6.7 Is maximum effort put into not needing to use attributive data?
☐ yes ☐ no
If yes, how? If not, why not?

7 **Process Control and Process Improvement:**
Comment: These questions examine whether the most effective method has been used for data collection. It is also important that the correct control chart is used. The reason for this is to ensure that, with the help of SPC, the right kind of action is taken in reaction to the data, and that these activities lead to permanent improvements.

7.1 Are control charts used and displayed on the shop floor?
☐ yes ☐ no
If yes, how? If not, why not?

7.2 How often are control charts used? Give a short explanation.

- ☐ Continually
- ☐ Weekly
- ☐ Monthly

7.3 Who interprets the control charts?
Check the relevant answers and give a short description of the way in which action is taken.

- ☐ Operator
- ☐ Process engineer
- ☐ SPC team
- ☐ Other

7.4 Which percentage of your data are measured continuously?

- ☐ 0%
- ☐ 1–25%
- ☐ 26–50%
- ☐ 51–75%
- ☐ 76–100%

7.5 What type of control chart is used for continuous data?
Check the relevant answers and give a short explanation.

- ☐ $(\overline{X}, R)$ chart
- ☐ $(\overline{X}, s)$ chart
- ☐ $(\overline{X}, MR)$ chart
- ☐ Other

7.6 What indication rules are used with the control charts?
Check the relevant answers and give a short explanation.

- ☐ Points outside the control limits
- ☐ Systematic deviations (*e.g.*, 8 points under the central line)

- ☐ Trends (successively increasing or decreasing points)
- ☐ Other rules

7.7 How often are notes made on the control chart or in the logbook concerning the actions taken after an out of control signal? Give a short explanation.

- ☐ Rarely
- ☐ Sometimes
- ☐ Often
- ☐ Always

7.8 What percentage of the data can be categorized as attributive?

- ☐ 0%
- ☐ 1–25%
- ☐ 26–50%
- ☐ 51–75%
- ☐ 76–100%

7.9 What types of attributive control charts are used?
Check the relevant answer and give a short explanation.

- ☐ p-chart
- ☐ c-chart
- ☐ Other

7.10 Which additional techniques are used to chart the degree to which a process is statistically controlled?
Check the relevant answer and give a short explanation.

- ☐ Pre-control techniques
- ☐ Trend diagrams
- ☐ Other

7.11 Which additional statistical techniques are used when solving problems?
Check the relevant answers and give a short explanation.

- ☐ Histograms
- ☐ Scatter diagrams
- ☐ Design of experiments
- ☐ Taguchi methods
- ☐ Pareto analyses
- ☐ Other

8 **Involvement of the Suppliers:**
Comment: Incoming goods and raw materials play an important role in the company's own process management. With the implementation of SPC the time will arise when the company's suppliers will need to become involved in the approach.

8.1 Which requirement do you dictate to your suppliers with regard to SPC?
Give a short explanation.

- ☐ None
- ☐ Voluntary
- ☐ Compulsory

8.2 What percentage of your suppliers has started on SPC?

- ☐ 0 %
- ☐ 1–25%
- ☐ 26–50%
- ☐ 51–75%
- ☐ 76–100%

8.3 How do you check the manner in which your suppliers implement SPC?
Check the relevant answers and give a short explanation.

- ☐ Supplied data
- ☐ Audits
- ☐ By means of an independent agency
- ☐ Other

9 **Total Judgement of the SPC Self-Evaluation:**
Comment: The company is asked to make a judgement on the state of the implementation on the basis of this self-evaluation. An indication can be given of the plans for the coming year.

9.1 Is it known which process steps must be controlled with the help of SPC techniques?
☐ yes ☐ no
If yes, how? If not, why not?

9.2 What percentage of the process steps has already been brought under the SPC method?

- ☐ 0%
- ☐ 1–25%
- ☐ 26–60%

- ☐ 51–75%
- ☐ 76–100%

N.B. For this question a minimum percentage of 70% must be attained in order to be able to state that the company has made reasonable progress with SPC. There should be a blueprint as to how 100% can ultimately be realized.

9.3 On the basis of this SPC self-evaluation, what is your judgement of the extent to which your company has made progress with SPC implementation?

- ☐ 0%
- ☐ 1–25%
- ☐ 26–50%
- ☐ 51–75%
- ☐ 76–100%

9.4 On the basis of your own judgement do you think that your company will shortly be ready for an SPC certification audit?

- ☐ Yes: when?
- ☐ No: what are your plans for the coming year?

References

[1] Al-Salti M., E.M. Aspinwall and A. Statham (1992), 'Implementing SPC in a Low Volume Manufacturing Environment', *Quality Forum*, **18(3)**, pp. 125–132.

[2] Amin R.W. and R.A. Ethridge (1998), 'A note on individual and moving range control charts', *Journal of Quality Technology*, **30**, pp. 70–74.

[3] Atkinson A.C. and H.-M. Mulira (1993), 'The stalactite plot for the detection of multivariate outliers', *Statistics and Computing*, **3**, pp. 27–35.

[4] Automotive Electronics Council (1994), *Quality System Assessment for Semiconductors Suppliers*, Michigan.

[5] Banks, D. (1993), 'Is industrial statistics out of control?', *Statistical Science*, **8**, pp. 356–409.

[6] Bhote, K.R. (1991), *World Class Quality*, Amacom, New York.

[7] Bissel, A.F. (1990), 'How reliable is your capability index,', *Applied Statistics*, **39**, pp. 331–340.

[8] Blonk, J. (1995), 'BPR: a responsibility of the shop floor', (in Dutch), *Sigma*, **95/3**, pp. 11–13.

[9] Boardman, T.J. (1994), 'The statistician who changed the world: W. Edwards Deming', 1900–1993, *American Statistician*, **48**, pp. 179–187.

[10] Box, G.E.P., W.G. Hunter and J.S. Hunter (1978), *Statistics for Experimenters*, Wiley, New York.

[11] Bureau of Business Practice (1992), *ISO 9000, Handbook of Quality Standards and Compliance*, Prentice Hall, Waterford.

[12] Camp, B.H. (1922), 'A new generalization of Tchebycheff's statistical inequality', *Bulletin of the American Mathematical Society*, **28**, pp. 427–432.

[13] Champ, C.W. and W.H. Woodall (1987), 'Exact results for Shewhart control charts with supplementary run rules', *Technometrics*, **29**, pp. 393–399.

[14] Chatfield, C. and A.J. Collins (1980), *Introduction to Multivariate Analysis*, Chapman and Hall, London.

[15] Chrysler, Ford and GM (1990), *Measurement System Evaluation, Reference Manual*, AIAG, Troy, Michigan.

[16] Cryer J.D. and T.P. Ryan (1990), 'The estimation of sigma for an X chart: $\overline{MR}/d_2$ or S/c_4?', *Journal of Quality Technology*, **22**, pp. 187–192.

[17] Davis, R.B. and W.H. Woodall (1988), 'Peformance of the control chart trend rule under linear shift', *Journal of Quality Technology*, **20**, pp. 260–262.

[18] Deming, W.E. (1986), *Out of the Crisis*, MIT, Cambridge, Massachusetts.

[19] Does, R.J.M.M. (1992), *Industrial statistics guarantees quality*, (in Dutch), Kwantitatieve Methoden 40, pp. 129–141.

[20] Does, R.J.M.M., J.A. van Dooremalen and R. Stam (1995a), *Statistical Process Control: The ASM Lithography Approach*, ASM Lithography, Veldhoven, The Netherlands.

[21] Does, R.J.M.M., J.A. van Dooremalen en R. Stam (1995b), *Statistische procesbeheersing (SPC): de ASM Lithography benadering*, ASM Lithography, Veldhoven, The Netherlands.

[22] Does, R.J.M.M. and K.C.B. Roes (1996), 'Industrial statistics and its recent contribution to total quality in the Netherlands', *Statistica Neerlandica*, **50**, pp. 27–51.

[23] Does, R.J.M.M., K.C.B. Roes and A. Trip (1995), 'A general set up for designing control charts in a mixed model', in: *The Proceedings of the 6th Seminar of the EOQ Committee on Statistical Methods, Budapest*, European Organization for Quality, Bern, Switzerland, pp. 58–71.

[24] Does, R.J.M.M. and B.F. Schriever (1992), 'Variables control chart limits and tests for special causes', *Statistica Neerlandica*, **46**, pp. 229–246.

[25] Does R.J.M.M., K.C.B. Roes and A. Trip (1998), 'Handling multivariate problems with univariate control charts', submitted for publication in *Journal of Chemometrics*.

[26] Duncan A.J. (1986), *Quality Control and Industrial Statistics*, Irwin, Homewood, Illinois.

[27] Engel J. and B. De Vries (1997), 'Evaluating a well known criterion for measurement precision', *Journal of Quality Technology*, **29**, pp. 469–476.

[28] Farnum N.R. (1992), 'Control Charts for Short Runs: Nonconstant Process and Measurement Error', *Journal of Quality Technology*, **24**, pp. 138–144.

[29] Farnum N.R (1995), 'Discussion: Q charts for Short Runs', *Journal of Quality Technology*, **27**, pp. 316–332.

[30] Gunter, B.H. (1989), 'Use and abuse of C_{pk}', *Quality Progress*, **23(1)**, pp. 72–73; **23(3)**, pp. 109; **23(5)**, pp. 79–80; **23(7)**, pp. 86–87.

[31] Haak A.T. (1994), *Dutch Sociotechnical Design in Practice: An Empirical Study of the Concept of the Whole Task Group*, Van Gorcum, Assen, The Netherlands.

[32] Hammer, M. and J. Champy (1993), *Re-Engineering the Corporation: A Manifesto for Business Revolution*, Harper Collins, New York.

[33] Harding A.J., K.R. Lee, J.L. Mullins (1992), *The Effect of Instabilities on Estimates of Sigma, 46th Annual Quality Congress Transactions*, ASQC, Milwaukee, WI, pp. 1037–1043.

[34] Hardjono, T.W. and F.W. Hes (1994), *The Dutch Quality Prize and Reward*, (in Dutch), Kluwer Bedrijfswetenschappen, Deventer, The Netherlands.

[35] Hare, L.B., R.W. Hoerl, J.D. Hromi and R.D. Snee (1995), 'The role of statistical thinking in management', *Quality Progress*, **29(2)**, pp. 53–60.

[36] Hauser, J.R. and D. Clausing (1988), 'The house of quality', *Harvard Business Review*, **66(3)**, pp. 63–73.

[37] Hillier F.S. (1964), '$\overline{X}$ chart control limits based on a small number of subgroups', *Industrial Quality Control*, **20**, pp. 24–29.

[38] Hillier F.S. (1967), 'Small sample probability limits for the range chart', *Journal of the American Statistical Association*, **62**, pp. 1488–1493.

[39] Hillier F.S. (1969), '$\overline{X}$- and R-chart control limits based on a small number of subgroups', *Journal of Quality Technology*, **1**, pp. 17–26.

[40] Hoerl, R. (1995), 'Enhancing the bottom line impact of statistical methods', *ASQC Statistics Division Newsletter*, **15(2)**, pp. 6–18.

[41] Imai, M. (1986), *Kaizen*, Random House, New York.

[42] Ishikawa, K. (1974), *Guide to Quality Control*, Asian Productivity Organization, Tokyo.

[43] Ishikawa, K. (1985), *What is Total Quality Control? The Japanese Way*, Prentice Hall, Englewood Cliffs, New Jersey.

[44] Joiner, B.L. (1994), *Fourth Generation Management: the New Business Consciousness*, McGraw Hill, New York.

[45] Jonkers B.S. (1997), *On $Q(R)$: Limitations and enhancements to $Q(R)$ control charts for the start-up phase of a production cycle*, MSc thesis, University of Amsterdam.

[46] Juran, J.M. (1989), *Leadership for Quality—an Executive Handbook*, Free Press, New York.

[47] Kamat A.R. (1953), 'On the Mean Successive Difference and its Ratio to the Root Mean Square', *Biometrika*, **40**, pp. 116–127.

[48] Kane, V.E. (1986), 'Process capability indices', *Journal of Quality Technology*, **18**, pp. 41–52 (corrigenda pp. 265).

[49] Laviolette M., J.W. Seaman Jr., J.D. Barrett and W.H. Woodall (1995), 'A probabilistic and statistical view of fuzzy methods (with discussion)', *Technometrics*, **37**, pp. 249–292.

[50] Ledolter J. and A. Swersey (1997), 'An evaluation of pre-control', *Journal of Quality Technology*, **29**, pp. 163–171.

[51] Lehmann E.L. (1986), *Testing Statistical Hypotheses*, 2nd edn, John Wiley & Sons, New York.

[52] Lorenzen, T.J. and L.C. Vance (1986), 'The economic design of control charts: A unified approach', *Technometrics*, **28**, pp. 3–10.

[53] Manly, B.F.J. (1986), *Multivariate Statistical Methods*, Chapman and Hall, Londen.

[54] Mason R.L., C.W. Champ, N.D. Tracy, S.J. Wierda and J.C. Young (1997), 'Assessment of multivariate process control techniques', *Journal of Quality Technology*, **29**, pp.140–143.

[55] Meidell, M.B. (1922), 'Sur un problème du calcul des probabilités et les statistiques mathématiques', *Comptes Rendus*, **175**, pp. 806–808.

[56] Montgomery, D.C. (1991), *Design and Analysis of Experiments*, Third edition, Wiley, New York.

[57] Montgomery, D.C. (1996), *Introduction to Statistical Quality Control*, Third edition, Wiley, New York.

[58] Myers R.H. and D.C. Montgomery (1997), 'A tutorial on generalised linear models', *Journal of Quality Technology*, **29**, pp. 274–291.

[59] Neave, H.R. (1990), *The Deming Dimension*, SPC Press, Knoxville, Tennessee.

[60] Neave H.R. (1992), *Why SPC?*, SPC Press, Knoxville, Tennessee.

[61] Nelson L.S. (1982), 'Control charts for individual measurements', *Journal of Quality Technology*, **14**, pp. 172–173.

[62] Nelson, L.S. (1984), 'The Shewhart control chart—Tests for special causes', *Journal of Quality Technology*, **16**, pp. 237–239.

[63] Nelson, L.S. (1985), 'Interpreting Shewhart control charts', *Journal of Quality Technology*, **17**, pp. 114–116.

[64] Nelson L.S. (1988), 'Control charts: Rational subgroups and effective applications', *Journal of Quality Technology*, **20**, pp. 7375.

[65] Nolan, T.W. and L.P. Provost (1990), 'Understanding variation', *Quality Progress*, vol 24(5), pp. 70–78.

[66] Palm, A.C. (1992), 'Some aspects of sampling for control charts', *ASQC Statistics Division Newletter*, **20**, pp. 73–75.

[67] Quesenberry C.P. (1991a), 'SPC Q charts for Start Up Processes and Short or Long Runs', *Journal of Quality Technology*, **23**, pp. 213–224.

[68] Quesenberry C.P. (1991b). 'SPC Q charts for a Binomial Parameter p: Short or Long Runs', *Journal of Quality Technology*, **23**, pp. 239–246.

[69] Quesenberry C.P. (1991c). 'SPC Q charts for a Poisson Parameter λ: Short or Long Runs', *Journal of Quality Technology*, **23**, pp. 296–303.

[70] Quesenberry, C.P. (1993), 'The effect of sample size on estimated limits for X? and X control charts', *Journal of Quality Technology*, **25**, pp. 237–247.

[71] Quesenberry, C.P. (1995), 'On properties of Q chart for variables', *Journal of Quality Technology*, **27**, pp. 184–203.

[72] Quesenberry C.P. (1995), 'Geometric Q charts for High Quality Processes (with discussion)', *Journal of Quality Technology*, **27**, pp. 304–333.

[73] Roes, K.C.B. and R.J.M.M. Does (1995), 'Shewhart type charts in nonstandard situations (with discussion)', *Technometrics*, **37**, pp. 15–40.

[74] Roes, K.C.B., R.J.M.M. Does and Y. Schurink (1993), 'Shewhart type control charts for individual observations', *Journal of Quality Technology*, **25**, pp. 188–198.

[75] Roes K.C.B., R.J.M.M. Does and B.S. Jonkers (1998), 'Effective application of $Q(R)$ charts in low volume manufacturing', accepted for publication, in: *Quality and Reliability Engineering International.*

[76] Sandorf, J.P. and A.T. Bassett III (1993), 'The OCAP: predetermined responses to out of control conditions', *Quality Progress*, **27(5)**, pp. 91–96.

[77] Satterthwaite F.E. (1954), *A simple, effective process control method*, Report 54-1, Rath&Strong Inc., Boston.

[78] Seder, L.A. (1990), 'Diagnosis with diagrams, Quality Engineering 2', pp. 505–530, (Reprinted from *Industrial Quality Control*, 1950).

[79] Shainin D. (1984), 'Better than good old $\overline{X}$ and R charts asked by vendees', *ASQC Quality Congress Transactions*, ASQC, Milwaukee, WI, pp. 302–307.

[80] Shewhart, W.A. (1931), *Economic Control of Quality of Manufactured Product*, Van Nostrand Reinhold, Princeton.

[81] Snee, R.D. (1990), 'Statistical thinking and its contribution to total quality', *American Statistician*, **44**, pp. 116–121.

[82] Snee, R.D. (1991), 'Can statisticians meet the challenge of total quality?', *Quality Progress*, **25(1)**, pp. 60–64.

[83] Stamatis, D.H. (1995), *Failure Mode and Effect Analysis: FMEA from Theory to Execution*, ASQC, Milwaukee, Wisconsin.

[84] Sullivan J.H., W.H. Woodall and M.M. Gardner (1995), Discussion on 'Shewhart type charts in nonstandard situations', *Technometrics*, **37**, pp. 31–35.

[85] Suzaki, K. (1987), *The New Manufacturing Challenge*, Free Press, New York.

[86] Suzaki, K. (1993), *The New Shop Floor Management*, Free Press, New York.

[87] Taguchi, G. (1986), *Introduction to Quality Engineering*, American Supplier Institute, Dearborn, Michigan.

[88] Veld, J. in 't (1988), *Analysis of Organizational Problems: an Application of Thinking in Systems and Processes*, (in Dutch), Fifth edition, Stenfert Kroese, Leiden, The Netherlands.

[89] Walker, E., J.W. Philpot and J. Clement (1991), 'False signal rates for the Shewhart control chart with supplementary runs tests', *Journal of Quality Technology*, **23**, pp. 247–252.

[90] Wetherill G.B. and D.W. Brown (1991), *Statistical Process Control; Theory and Practice*, Chapman and Hall, London.

[91] Wheeler, D.J. (1991), *Short Run SPC*, SPC Press, Knoxville, Tennessee.

[92] Wheeler, D.J. and D.S. Chambers (1992), *Understanding Statistical Process Control*, Second edition, SPC Press, Knoxville Tennessee.

[93] Wierda S.J. (1994), *Multivariate Statistical Process Control*, PhD thesis, Wolters–Noordhoff, Groningen, The Netherlands.

[94] Woodall, W.H. (1986), 'Weaknesses of the economic design of control charts', Letter to the editor, *Technometrics*, **28**, pp. 408–409.

[95] Woodall W.H. (1997), 'Control charts based on attributive data: Bibliography and review', *Journal of Quality Technology*, **29**, pp. 172–183.

Index

MATHEMATICAL MODELLING:
Theory and Applications

1. M. Křížek and P. Neittaanmäki: *Mathematical and Numerical Modelling in Electrical Engineering*. Theory and Applications. 1996
ISBN 0-7923-4249-6
2. M.A. van Wyk and W.-H. Steeb: *Chaos in Electronics*. 1997
ISBN 0-7923-4576-2
3. A. Halanay and J. Samuel: *Differential Equations, Discrete Systems and Control*. Economic Models. 1997 ISBN 0-7923-4675-0
4. N. Meskens and M. Roubens (eds.): *Advances in Decision Analysis*. 1999
ISBN 0-7923-5563-6
5. R.J.M.M. Does, K.C.B. Roes and A. Trip: *Statistical Process Control in Industry*. Implementation and Assurance of SPC. 1999
ISBN 0-7923-5570-9

KLUWER ACADEMIC PUBLISHERS – DORDRECHT / BOSTON / LONDON

MIX
Papier aus verantwortungsvollen Quellen
Paper from responsible sources
FSC® C105338

If you have any concerns about our products, you can contact us on
ProductSafety@springernature.com

In case Publisher is established outside the EU, the EU authorized representative is:
Springer Nature Customer Service Center GmbH
Europaplatz 3, 69115 Heidelberg, Germany

Printed by Libri Plureos GmbH
in Hamburg, Germany